D0897377

A Guide to
IUPAC Nomenclature
of Organic Compounds
Recommendations 1993

International Union of Pure and Applied Chemistry
Organic Chemistry Division
Commission on Nomenclature of Organic Chemistry (III.1)

A Guide to
IUPAC Nomenclature
of Organic Compounds
Recommendations 1993

(including revisions, published and hitherto unpublished, to the
1979 edition of *Nomenclature of Organic Chemistry*)

Prepared for publication by
R. PANICO, W. H. POWELL and
JEAN-CLAUDE RICHER (Senior Editor)

OXFORD
BLACKWELL SCIENTIFIC PUBLICATIONS
LONDON EDINBURGH BOSTON
MELBOURNE PARIS BERLIN VIENNA

© 1993 International Union of Pure and
Applied Chemistry and published for them by
Blackwell Scientific Publications
Editorial Offices:
Osney Mead, Oxford OX2 0EL
25 John Street, London WC1N 2BL
23 Ainslie Place, Edinburgh EH3 6AJ
238 Main Street, Cambridge
  Massachusetts 02142, USA
54 University Street, Carlton
  Victoria 3053, Australia

Other Editorial Offices:
Librairie Arnette SA
1, rue de Lille
75007 Paris
France

Blackwell Wissenschafts-Verlag GmbH
Düsseldorfer Str. 38
D-10707 Berlin
Germany

Blackwell MZV
Feldgasse 13
A-1238 Wien
Austria

SECTIONS A AND B
First edition 1958, Second edition 1966
Third edition 1971 (combined with Section C)
Fourth edition 1979 (combined with Sections C,
  D, E, F & H)

SECTION C
First edition 1965
Second edition 1971 (combined with Sections A & B)
Third edition 1979 (combined with Sections A,
  B, D, E, F & H)

SECTIONS D, E, F, H
First edition 1979 (combined with Sections A, B & C)

This edition published 1993

The contents of the first edition of Section C appeared
in *Pure and Applied Chemistry*, Vol. 11, Nos 1–2 (1965).
The contents of the first edition of Section E appeared
in *Pure and Applied Chemistry*, Vol. 45, No. 1 (1976)
and of the first edition of Section H
in Vol. 51, No. 2 (1979)

Set by Macmillan India Ltd, Bangalore-25
Printed and bound in Great Britain
at the University Press, Cambridge

DISTRIBUTORS

Marston Book Services Ltd
PO Box 87
Oxford OX2 0DT
(*Orders*: Tel: 0865 791155
       Fax: 0865 791927
       Telex: 837515)

Australia
Blackwell Scientific Publications Pty Ltd
54 University Street
Carlton, Victoria 3053
(*Orders*: Tel: (03) 347–5552)

USA and North America
CRC Press, Inc.
2000 Corporate Blvd, NW
Boca Raton
Florida 33431

A catalogue record for this book
is available from the British Library

ISBN 0-632-03488-2

Library of Congress
Cataloging-in-Publication Data

International Union of Pure and Applied Chemistry.
  Commission on the Nomenclature
  of Organic Chemistry.
  A guide to IUPAC nomenclature
  of organic compounds:
  recommendations 1993/International Union
  of Pure and Applied Chemistry,
  Organic Chemistry Division,
  Commission on Nomenclature of
  Organic Chemistry (III.1);
  prepared for publication by R. Panico and
  W. H. Powell; Jean-Claude Richer, senior editor.
      p.   cm.
  "Version: 92.08.15."
  Includes bibliographical references
  and index.
  ISBN 0-632-03488-2
  1. Chemistry, Organic—Nomenclature.
  I. Panico, Robert, 1925–   . II. Powell, W. H.
  III. Richer, Jean-Claude.   IV. Title.
  QD291.157   1993
  547′.0014—dc20

Contents

Membership of the Commission during the Preparation of this Guide, xi

List of Tables, xii

Preamble, xiii

R-0	INTRODUCTION, 1
R-0.0	Scope, 1
R-0.1	Conventions, 1
R-0.1.1	Spelling, 1
R-0.1.2	Position of locants, 1
R-0.1.3	Punctuation, 2
R-0.1.3.1	Commas, 2
R-0.1.3.2	Full stops, 2
R-0.1.3.3	Colons, 2
R-0.1.3.4	Hyphens, 2
R-0.1.3.5	Spaces, 3
R-0.1.4	Numerical (multiplicative) prefixes, 4
R-0.1.5	Enclosing marks, 5
R-0.1.5.1	Parentheses, 5
R-0.1.5.2	Square brackets, 6
R-0.1.5.3	Braces, 7
R-0.1.6	Italicization, 7
R-0.1.6.1	Lower case italic letters, 8
R-0.1.6.2	Italicized element symbols, 8
R-0.1.6.3	Italic words, syllables and capital roman letters, 8
R-0.1.7	Elision and addition of vowels, 8
R-0.1.8	Order of prefixes, 10
R-0.2	Glossary, 13
R-0.2.1	Parent structures, 13
R-0.2.1.1	Parent hydride, 13
R-0.2.1.2	Functional parent, 13
R-0.2.2	Groups, 13
R-0.2.2.1	Substituent atom or group, 13
R-0.2.2.2	Characteristic group, 13
R-0.2.2.3	Principal group, 13
R-0.2.3	Names, 14
R-0.2.3.1	Trivial name, 14
R-0.2.3.2	Semisystematic name or semitrivial name, 14
R-0.2.3.3	IUPAC name, 14
R-0.2.3.3.1	Fusion name, 14
R-0.2.3.3.2	Hantzsch–Widman name, 14
R-0.2.3.3.3	Functional class name, 14
R-0.2.3.3.4	Radicofunctional name, 15
R-0.2.3.3.5	Replacement name, 15
R-0.2.3.3.6	Substitutive name, 15

CONTENTS

R-0.2.3.3.7   Conjunctive name, 15
R-0.2.3.3.8   Additive name, 16
R-0.2.3.3.9   Subtractive name, 16
R-0.2.3.3.10   Multiplicative name, 17
R-0.2.4   Other terms used in these recommendations, 17
R-0.2.4.1   Seniority, senior, 17
R-0.2.4.2   Lowest set of locants, 17

R-1   GENERAL PRINCIPLES OF ORGANIC NOMENCLATURE, 18
R-1.0   Introduction, 18
R-1.1   Bonding number, 20
R-1.1.1   Definition, 21
R-1.1.2   Standard bonding numbers, 21
R-1.1.3   Nonstandard bonding numbers, 21
R-1.1.4   Contiguous formal double bonds, 21
R-1.2   Nomenclature operations, 22
R-1.2.1   Substitutive operation, 22
R-1.2.2   Replacement operation, 23
R-1.2.3   Additive operation, 24
R-1.2.3.1   By use of an additive prefix, 24
R-1.2.3.2   By use of an additive suffix, 25
R-1.2.3.3   By use of a separate word, 25
R-1.2.3.4   By connecting the names of the components of an addition compound with a dash (long hyphen), 26
R-1.2.3.5   By juxtaposition or multiplication of substituent prefix terms, 26
R-1.2.4   Conjunctive operation, 26
R-1.2.4.1   By juxtaposition of component names, 26
R-1.2.4.2   By placing a multiplicative prefix before the name of the corresponding parent hydride, 27
R-1.2.5   Subtractive operation, 27
R-1.2.5.1   By use of a prefix, 28
R-1.2.5.2   By a change in ending or addition of a suffix, 29
R-1.2.6   Ring formation or cleavage, 30
R-1.2.6.1   The prefix cyclo-, 30
R-1.2.6.2   The prefix seco-, 31
R-1.2.7   Rearrangement, 31
R-1.2.7.1   The prefix *abeo-*, 31
R-1.2.7.2   The prefix *retro-*, 32
R-1.2.8   Multiplicative operation, 32
R-1.3   Indicated hydrogen, 34

R-2   PARENT HYDRIDES AND THEIR DERIVED SUBSTITUENT GROUPS, 36
R-2.0   Introduction, 36
R-2.1   Mononuclear hydrides, 36
R-2.2   Acyclic polynuclear hydrides, 36
R-2.2.1   Hydrocarbons, 36
R-2.2.2   Homogeneous hydrides other than hydrocarbons or boron hydrides, 37
R-2.2.3   Heterogeneous hydrides, 38
R-2.3   Monocyclic hydrides, 39
R-2.3.1   Hydrocarbons, 39
R-2.3.2   Homogeneous hydrides other than hydrocarbons or boron hydrides, 39
R-2.3.3   Heterogeneous hydrides other than heteropolyboron hydrides, 40
R-2.4   Polycyclic parent hydrides, 44
R-2.4.1   Fusion nomenclature, 44
R-2.4.2   Bridged parent hydrides – extension of the von Baeyer system, 49
R-2.4.2.1   Bicyclic ring systems, 49

CONTENTS

R-2.4.2.2   Polycyclic ring systems, 50
R-2.4.3   Spiro parent hydrides, 51
R-2.4.3.1   Monospiro parent hydrides, 51
R-2.4.3.2   Polyspiro parent hydrides, 52
R-2.4.3.3   Spiro parent hydrides containing polycyclic ring systems, 53
R-2.4.4   Ring assemblies, 53
R-2.4.4.1   Assemblies of two identical cyclic systems, 53
R-2.4.4.2   Unbranched assemblies consisting of three or more identical ring systems, 54
R-2.4.5   Cyclophanes, 55
R-2.4.6   Natural product parent hydrides, 55
R-2.5   Substituent prefix names derived from parent hydrides, 56

R-3   **CHARACTERISTIC (FUNCTIONAL) GROUPS, 59**
R-3.0   Introduction, 59
R-3.1   Unsaturation, 59
R-3.1.1   Suffixes denoting multiple bonds, 59
R-3.1.2   Hydro prefixes, 60
R-3.1.3   Dehydro prefixes, 61
R-3.1.4   Substituent prefix names for unsaturated/saturated parent hydrides, 62
R-3.2   Specification of characteristic groups, 62
R-3.2.1   Prefixes and suffixes, 62
R-3.2.2   Functional modifiers, 64
R-3.3   Functional parent compounds and derived substituent groups, 65
R-3.4   Functional replacement, 66

R-4   **GUIDE TO NAME CONSTRUCTION, 68**
R-4.0   Introduction, 68
R-4.1   General principles, 68
R-4.2   Examples, 72

R-5   **APPLICATIONS TO SPECIFIC CLASSES OF COMPOUNDS, 77**
R-5.0   Introduction, 77
R-5.1   Binary hydrides and related parent hydrides, 77
R-5.1.1   Hydrocarbons, 77
R-5.1.2   Chalcogen hydrides, 78
R-5.1.3   Hydrides of the group 15 elements, 78
R-5.1.4   Silicon parent hydrides, 79
R-5.1.4.1   Silanes, 79
R-5.1.4.2   Heterogeneous silicon hydrides: siloxanes and analogues, 79
R-5.2   Organometallic compounds, 81
R-5.2.1   Organometallic compounds of antimony, bismuth, germanium, tin, and lead, 81
R-5.2.2   Organometallic compounds in which the metal is bound only to carbon atoms of organic groups and hydrogen, 81
R-5.2.3   Organometallic compounds with anionic ligands, 82
R-5.3   Halogen, nitro, nitroso, azo, diazo and azido compounds, 82
R-5.3.1   Halogen compounds, 82
R-5.3.2   Nitro and nitroso compounds, 83
R-5.3.3   Azo, azoxy, diazo, and related compounds, 83
R-5.3.3.0   Diazenes, 83
R-5.3.3.1   Azo compounds, 84
R-5.3.3.2   Azoxy compounds, 85
R-5.3.3.3   Diazonium compounds, 86
R-5.3.3.4   Azo compounds with the general structure $R-N=N-X$, 86
R-5.3.3.5   Diazo compounds, 87
R-5.3.4   Azides, 87
R-5.3.5   Isodiazenes, 87

CONTENTS

R-5.4   Amines and imines, 87
R-5.4.1   Primary amines, 88
R-5.4.2   Secondary and tertiary amines, 88
R-5.4.3   Imines, 89
R-5.4.4   Hydroxylamines, 90
R-5.4.5   Amine oxides, 90
R-5.5   Hydroxy compounds, their derivatives and analogues, 91
R-5.5.1   Hydroxy compounds and analogues, 91
R-5.5.1.1   Alcohols and phenols, 91
R-5.5.1.2   Sulfur, selenium, and tellurium analogues of alcohols and phenols, 92
R-5.5.2   Substituent prefixes derived from alcohols, phenols, and their analogues, 93
R-5.5.3   Salts, 94
R-5.5.4   Ethers and chalcogen analogues, 94
R-5.5.4.1   Substitutive names, 94
R-5.5.4.2   Functional class names, 95
R-5.5.4.3   Replacement nomenclature, 95
R-5.5.4.4   Cyclic ethers, 95
R-5.5.5   Hydroperoxides and peroxides, 96
R-5.5.6   Hydropolysulfides and polysulfides, 97
R-5.5.7   Sulfoxides, sulfones, and their analogues, 97
R-5.6   Aldehydes, ketones, their derivatives and analogues, 98
R-5.6.1   Aldehydes, thioaldehydes, and their analogues, 98
R-5.6.2   Ketones, thioketones, and their analogues, 100
R-5.6.2.1   Ketones, 100
R-5.6.2.2   Chalcogen analogues of ketones, 101
R-5.6.3   Ketenes, 101
R-5.6.4   Acetals, hemiacetals, acylals, and their analogues, 102
R-5.6.4.1   Acetals, 102
R-5.6.4.2   Hemiacetals, 103
R-5.6.4.3   Acylals, 103
R-5.6.5   Acyloins, 104
R-5.6.6   Nitrogenous derivatives of carbonyl compounds, 104
R-5.6.6.1   Oximes, 104
R-5.6.6.2   Hydrazones, 105
R-5.6.6.3   Azines, 105
R-5.6.6.4   Other nitrogen derivatives of carbonyl compounds, 105
R-5.7   Acids and related characteristic groups, 106
R-5.7.1   Carboxylic acids, 107
R-5.7.1.1   Simple (unsubstituted) acyclic mono- and dicarboxylic acids, 107
R-5.7.1.2   Substituted carboxylic acids, 109
R-5.7.1.2.1   Hydroxy, alkoxy, and oxo acids, 109
R-5.7.1.2.2   Amic and anilic acids, 109
R-5.7.1.2.3   Amino acids, 110
R-5.7.1.3   Modification of carboxylic acid suffixes, 110
R-5.7.1.3.1   Peroxy acids, 110
R-5.7.1.3.2   Imidic, hydrazonic, and hydroximic acids, 110
R-5.7.1.3.3   Hydroxamic acids, 112
R-5.7.1.3.4   Thiocarboxylic and thiocarbonic acids, 112
R-5.7.2   Chalcogen acids containing chalcogen atoms directly linked to an organic group, 113
R-5.7.2.1   Sulfur acids containing sulfur atoms directly linked to an organic group, 113
R-5.7.2.2   Selenium acids containing selenium directly linked to an organic group, 115
R-5.7.3   Phosphorus and arsenic acids containing phosphorus or arsenic atoms directly linked to an organic group, 115
R-5.7.3.1   Phosphorus oxo acids and replacement modifications, 115
R-5.7.3.2   Arsenic oxo acids and replacement modifications, 116
R-5.7.4   Salts and esters, 117

CONTENTS

R-5.7.4.1   Salts, 117
R-5.7.4.2   Esters, 117
R-5.7.5   Lactones, lactams, lactims, and analogues, 119
R-5.7.5.1   Lactones, 120
R-5.7.5.2   Sultones, 120
R-5.7.5.3   Lactams and lactims, 121
R-5.7.5.4   Sultams, 121
R-5.7.6   Acid halides, 122
R-5.7.7   Anhydrides and their analogues, 123
R-5.7.7.1   Symmetrical anhydrides, 123
R-5.7.7.2   Unsymmetrical (mixed) anhydrides, 124
R-5.7.7.3   Chalcogen analogues of anhydrides, 124
R-5.7.8   Amides, imides, and hydrazides, 125
R-5.7.8.1   Monoacyl derivatives of ammonia (primary amides), 125
R-5.7.8.2   Symmetrical diacyl and triacyl derivatives of ammonia, 127
R-5.7.8.3   Imides, 128
R-5.7.8.4   Hydrazides, 128
R-5.7.9   Nitriles, isocyanides and related compounds, 129
R-5.7.9.1   Nitriles, 129
R-5.7.9.2   Cyanide-related compounds, 131
R-5.7.9.3   Nitrile oxides, 131
R-5.8   Radicals and ions, 132
R-5.8.1   Radicals, 132
R-5.8.1.1   Monovalent radicals, 132
R-5.8.1.2   Divalent and trivalent radicals, 132
R-5.8.1.3   Radical centres on characteristic groups, 134
R-5.8.2   Cations, 136
R-5.8.3   Anions, 139
R-5.8.4   Cationic and anionic centres in a single structure, 141
R-5.8.5   Radical ions, 141

R-6   NAME INTERPRETATION, 143
R-6.0   Introduction, 143
R-6.1   6-(4-Hydroxyhex-1-en-1-yl)undeca-2,4-diene-7,9-diyne-1,11-diol, 143
R-6.2   2,3-Dichloro-6-[4-chloro-2-(hydroxymethyl)-5-oxohex-3-en-1-yl]pyridine-4-carboxylic acid, 144
R-6.3   3-(2,3-Dihydroxypropyl)-α-methylquinoline-2-pentanoic acid, 145
R-6.4   4,4′-Dinitro-2,3′-[ethylenebis(sulfanediyl)]dicyclohexane-1-carbaldehyde, 146
R-6.5   1-Methylbutyl 4-(2-acetyl-2-ethylhydrazino)benzoate, 147

R-7   STEREOCHEMICAL SPECIFICATION, 149
R-7.0   Introduction, 149
R-7.1   cis-trans Isomerism – the E/Z convention, 149
R-7.1.0   Introduction, 149
R-7.1.1   cis and trans Isomers, 149
R-7.1.2   The E/Z convention, 151
R-7.2   Chiral compounds – specification of absolute configuration, 152
R-7.2.1   The R/S convention, 152
R-7.2.2   Relative configuration, 154

R-8   ISOTOPICALLY MODIFIED COMPOUNDS, 155
R-8.0   Introduction, 155
R-8.1   Symbols and definitions, 155
R-8.1.1   Nuclide symbols, 155
R-8.1.2   Atomic symbols, 155
R-8.1.3   Isotopically unmodified compounds, 156
R-8.1.4   Isotopically modified compounds, 156

CONTENTS

R-8.2   Isotopically substituted compounds, 156
R-8.2.1   Formulae, 156
R-8.2.2   Names, 157
R-8.3   Isotopically labelled compounds, 157
R-8.3.1   Specifically labelled compounds, 157
R-8.3.2   Selectively labelled compounds, 158
R-8.3.3   Nonselectively labelled compounds, 160
R-8.3.4   Isotopically deficient compounds, 160

R-9   APPENDIX, 162
R-9.0   Introduction, 162
R-9.1   Trivial and semisystematic names retained for naming organic compounds, 162
R-9.2   Bridge names, 180
R-9.2.1   Simple bivalent bridges, 180
R-9.2.2   Simple polyvalent bridges, 182
R-9.3   'a' Prefixes used in replacement nomenclature, 182

INDEX, 183

Membership of the Commission during the Preparation of this Guide (1979–1991)

Titular members

O. Achmatowicz (Poland), 1979–1987; H. J. T. Bos (Netherlands), 1987–1991; J. R. Bull (Republic of South Africa), 1987–1991; H. A. Favre (Canada), 1989–1991; P. M. Giles, Jr. (USA), 1989–1991; E. W. Godly (UK), 1987–1991, *Secretary*, 1989–1991; D. Hellwinkel (Federal Republic of Germany), 1979–1987, *Vice-Chairman*, 1981–1987; K. Hirayama (Japan), 1975–1983; A. D. McNaught (UK), 1979–1987; G. P. Moss (UK), 1977–1987, *Chairman*, 1981–1987, *Vice-Chairman*, 1979–1981; R. Panico (France), 1981–1991, *Vice-Chairman*, 1989–1991; W. H. Powell (USA), *Secretary*, 1979–1989; J. C. Richer (Canada), 1979–1989, *Vice-Chairman*, 1987–1989; J. Rigaudy (France), 1967–1981, *Chairman*, 1977–1981; P. A. S. Smith (USA), 1983–1991, *Chairman*, 1987–1991; O. Weissbach (Federal Republic of Germany), 1987–1991.

Associate members

O. Achmatowicz (Poland), 1987–1989; K. Bláha* (Czechoslovakia), 1979–1987; H. J. T. Bos (Netherlands), 1983–1987; A. J. Boulton (UK), 1983–1987; J. R. Bull (Republic of South Africa), 1985–1987; L. C. Cross* (U.K.), 1977–1981; D. Eckroth (USA), 1975–1983; F. Fariña (Spain), 1989–1991; H. Favre (Canada), 1987–1989; J. H. Fletcher (USA), 1975–1983; P. M. Giles, Jr. (USA), 1983–1989; E. W. Godly (UK), 1979–1987; P. Grünanger (Italy), 1987–1991; H. Grünewald (Federal Republic of Germany), 1989–1991; H. Gutmann (Switzerland), 1983–1989; J. Heger (Czechoslovakia), 1985–1989; D. Hellwinkel (Federal Republic of Germany), 1987–1989; K. Hirayama (Japan), 1983–1987; R. J.-R. Hwu (USA), 1989–1991; M. A. C. Kaplan (Brazil), 1989–1991; M. V. Kisakürek (Switzerland), 1987–1991; S. P. Klesney (USA), 1979–1985; W. Liebscher (Federal Republic of Germany), 1989–1991; K. L. Loening (USA), 1979–1983; N. Lozac'h (France), 1977–1987; A. D. McNaught (UK), 1987–1989; M. M. Mikołajczyk (Poland), 1989–1991; G. P. Moss (UK), 1987–1989; R. Panico (France), 1979–1981; J. Rigaudy (France), 1981–1985; C. Schmitz (France), 1989–1991; R. Schoenfeld* (Australia), 1981–1987; P. A. S. Smith (USA), 1979–1983; D. Tavernier (Belgium), 1987–1991; J. G. Traynham (USA), 1989–1991; F. Vögtle (Federal Republic of Germany), 1972–1983; O. Weissbach (Federal Republic of Germany), 1979–1987.

National representatives

H. Y. Aboul Enein (Saudi Arabia), 1988–1989; O. Achmatowicz (Poland), 1989–1991; A. T. Balaban (Romania), 1983–1989; H. J. T. Bos (Netherlands), 1981–1983; J. R. Bull (Republic of South Africa), 1983–1985; J. R. Cannon (Australia), 1982–1987; K. C. Chan (Malaysia), 1983–1987; G. Deák** (Hungary), 1979–1991; F. Fariña (Spain), 1987–1989; M. J. Gasić (Yugoslavia), 1989–1991; P. Grünanger (Italy), 1984–1987; W.-Y. Huang (Peoples Republic of China), 1981–1987; S. Ikegami (Japan), 1986–1991; A. Ikizler (Turkey), 1987–1991; J. Kahovec (Czechoslovakia), 1989–1991; M. A. C. Kaplan (Brazil), 1983–1985; G. L'abbé (Belgium), 1981–1985; X. T. Liang (Peoples Republic of China), 1987–1991; L. Maat (Netherlands), 1989–1991; G. Mehta (India), 1983–1985; L. J. Porter (New Zealand), 1987–1991; J. A. Retamar (Argentina), 1980–1985; H. Schick (German Democratic Republic), 1987–1991; R. Schoenfeld* (Australia), 1980; S. Swaminathan (India), 1985–1991; D. Tavernier (Belgium), 1986–1987.

*Deceased
**Deceased 1992

List of Tables

1 Examples of nomenclature operations, 19
2 Mononuclear hydrides, 37
3 Hantzsch–Widman system prefixes, 42
4 Hantzsch–Widman system stems, 42
5 Suffixes and prefixes for some important characteristic groups in substitutive nomenclature, 63
6 Affixes for ionic and radical centres in parent structures, 64
7 Functional parent acids and derived substituent groups of nitrogen and phosphorus, 65
8 Functional replacement prefixes and infixes, 66
9 Characteristic groups cited only as prefixes in substitutive nomenclature, 69
10 General classes of compounds in decreasing order of priority for choosing and naming a principal characteristic group, 70
11 Basic numerical terms (multiplying affixes), 71
12 Suffixes and endings for carboxylic acids, some related characteristic groups, and substituted derivatives, 107
13 Suffixes for replacement analogues of carboxylic acids, 111
14 Suffixes for sulfur acids and replacement modifications, 114
15 Functional parent compounds for phosphorus acids and functional replacement modifications, 116
16 Cyanide and related groups in order of decreasing priority for citation as functional class name, 131
17 Mononuclear parent onium ions, 136
18 Comparative examples of formulae and names for isotopically modified compounds, 161
19 Acyclic and monocyclic hydrocarbons, 162
20 Unsaturated polycyclic hydrocarbons, 164
21 Saturated polycyclic hydrocarbons, 166
22 Polycyclic hydrocarbon substituent prefixes, 166
23 Heterocyclic parent hydrides, 166
24 Hydrogenated heterocyclic parent hydrides, 171
25 Heterocyclic substituent groups and cations, 172
26 Hydroxy compounds, ethers, and related substituent groups, 173
27 Carbonyl compounds and derived substituent groups, 174
28 Carboxylic acids and related groups, 175
29 Amines, nitrogenous heterocyclic parent structures, and derived substituent groups, 177
30 Sulfides, sulfonic acids, and derived substituent groups, 178
31 Acyclic polynitrogen parent structures and derived substituent groups, 179
32 Halogen compounds, 180
33 'a' Prefixes used in replacement nomenclature, 182

Preamble

The main purpose of chemical nomenclature is to identify a chemical species by means of written or spoken words. To be useful for *communication* among chemists, nomenclature for chemical compounds should additionally contain within itself an explicit or implied relationship to the structure of the compound, in order that the reader or listener can deduce the structure (and thus the identity) from the name. This purpose requires a system of principles and rules, the application of which gives rise to a *systematic nomenclature* (for examples, see the 1979 Edition of the IUPAC *Nomenclature of Organic Chemistry*[1]).

In contrast to such systematic names, there are traditional names, semisystematic or trivial, which are widely used for a core group of common compounds. Examples are 'acetic acid', 'benzene', 'cholesterol', 'styrene', 'formaldehyde', 'water', 'iron'. Many of these names are also part of general nonscientific language and are thus not confined to use within the science of chemistry. They are useful, and in many cases indispensable (consider the alternative systematic name for cholesterol, for example). Little is to be gained, and certainly much to be lost, by replacing such names. Therefore, where they meet the requirements of utility and precision, and can be expected to continue to be widely used by chemists and others, they are retained and, for the most part, preferred in this Guide.

Semisystematic names also exist, such as 'methane', 'propanol', and 'benzoic acid', which are so familiar that few chemists realize that they are not fully systematic. They are retained, and indeed, in some cases there are no better systematic alternatives.

It is important to recognize that the rules of systematic nomenclature need not necessarily lead to a unique name for each compound, but must always lead to an unambiguous one. Lucidity in communication often requires that the rules be applied with different priorities. A comparative discussion of the compounds $CH_3-CH=CH_2$, $Cl-CH_2-CH=CH_2$, $C_6H_5-CH_2-CH=CH_2$, $H_2N-CH_2-CH=CH_2$ and $HO-CH_2-CH=CH_2$, might be easier to follow if they were all named as propenes, even though with the last three, the benzene ring, amino, and hydroxy groups may have seniority over the double bond for citation as a parent or as a suffix. In other cases, a set of rules that generates clear and efficient names for some compounds can lead to clumsy and nearly unrecognizable names for others, even closely related ones. To force the naming of all compounds into the Procrustean bed of one set of rules would not serve the needs of general communication, and the Commission believes that the majority of organic chemists would not accept such a policy for general communication. This situation can be illustrated by a compound that most chemists would probably name as 'pentaphenylethane', instinctively, whereas the application of a principle favouring rings

---

[1] International Union of Pure and Applied Chemistry, Organic Chemistry Division, Commission on Nomenclature of Organic Chemistry, *Nomenclature of Organic Chemistry, Sections A, B, C, D, E, F, and H*, 1979 Edition, J. Rigaudy and S. P. Klesney, eds, Pergamon Press, Oxford, 1979, 559 pp.

over chains leads to a name such as 'ethanepentaylpentabenzene'. The first name is certainly more easily recognized than the second. Another example is compound (1), which would be named 'benzoazete' by application of fusion nomenclature principles, but von Baeyer nomenclature would give the name '7-azabicyclo[4.2.0]octa-1,3,5,7-tetraene'. On the other hand, von Baeyer nomenclature is most useful with compounds such as 'bicyclo[4.3.2]undecane' (2).

(1)   (2)

In view of the foregoing considerations, this *Guide to IUPAC Nomenclature of Organic Compounds* often presents alternative sets of rules, equally systematic, wherever available and justifiable, to enable a user to fit the name to a particular need.

Lastly, the Commission recognizes that for certain types of compounds, there is significant disagreement among chemists in different fields as to what should be the preferred nomenclature. This situation leads to an apparent lack of decisiveness in some of the recommendations in this document. This is unavoidable, because long experience has taught that formulating rules not having general support is a futile exercise; such rules will be widely ignored. Therefore, the Commission's policy is to offer critically examined alternatives, some of which may be new proposals, and to observe how they are accepted and used. If one of the alternatives subsequently becomes preferred to an overwhelming extent by the community of chemists, a future edition of recommendations can reflect that fact.

In this Guide, some practices of the Chemical Abstracts Service and/or of the Beilstein Institute have been mentioned. This is done only for informational purposes, and such instances are not necessarily recommendations of the Commission. The Commission recognized that there are circumstances that require a preferred (i.e., unique) name. These include comprehensive indexing (such as for the volume indexes to *Chemical Abstracts*) in order to avoid an intolerable amount of cross-indexing and multiple entries. This need is being met in a particular way by Chemical Abstracts Service as in-house procedures designed to place compounds with the same parent skeleton together while at the same time minimizing the number of rules. The *Chemical Abstracts Index Guide* treats the majority of compounds, but is not complete. There are a number of other in-house procedures applied elsewhere, such as in *Beilstein*[2] (not yet explicitly published). Specialty files, such as steroids and carbohydrates, have different bases. The result is not always compatible with ease of recognition, ease of generation, or conciseness. Unique names are also very important in legal situations, with manifestations in patents, export-import regulations, health and safety information, etc. These needs are being addressed by other Commission projects. In this Guide, the primary aim is to provide directions for arriving at an unambiguous name, although some guidance is given about establishing preferences (see, for example, Tables 10 and 16).

The rules given in the *Nomenclature of Organic Chemistry*[1], commonly known as the 'Blue Book', emphasize the generation of unambiguous names in accord with the

---

[2] *Beilsteins Handbuch der Organischen Chemie*, Springer–Verlag, Berlin.

historical development of the subject, because the need for a 'unique' name was not perceived to be compelling by earlier generations of chemists. The so-called information explosion of recent decades is a major factor in changing this perception. The present matrix of rules, however, cannot easily be overlaid with a simple set of principles for selecting a preferred name among the systematic alternatives, and to declare a preference by arbitrary fiat in each situation would surely lead to widespread rejection. Therefore, the Commission has initiated projects to formulate a comprehensive guide for selecting unique names that will, insofar as feasible, have good recognition value and a general acceptance among chemists; the results will be presented in later publications. Further projects, with longer-range objectives of systematizing nomenclature of organic compounds, are also under way.

Finally, those who use this Guide should be aware that the nomenclature in these recommendations is independent of the orientation of the graphic structure (except as stated in the recommendations for nomenclature of fused-ring structures) and of conformation. Furthermore, the nomenclature does not indicate nor imply an electronic structure or spin multiplicity.

R-0   Introduction

R-0.0   SCOPE

This guide to IUPAC nomenclature of organic compounds provides an outline of the main principles of organic nomenclature as described in the 1979 edition of the IUPAC *Nomenclature of Organic Chemistry*[1] and includes important changes agreed upon since its publication. Differences from the 1979 edition have not been specifically highlighted. However, in many cases a name used in the 1979 edition, preceded by a word such as 'formerly' or 'previously', appears in parentheses following a name recommended herein.

R-0.1   CONVENTIONS

In this guide, efforts have been made to systematize the style (spelling, position of locants, typography, punctuation, italicization, etc.) of the names of organic compounds according to the IUPAC English style. As usual, IUPAC recognizes the needs of other languages to introduce their own modifications. Indeed, even for the English language these conventions do not have the status of recommendations because the Commission recognizes that differences appear in names found in the *Beilsteins Handbuch der Organischen Chemie* and in the volume indexes to *Chemical Abstracts*, both of which have more stringent requirements than are necessary for the general literature.

R-0.1.1   **Spelling**

The spelling of elements is that given in the IUPAC *Nomenclature of Inorganic Chemistry*[3]; for example, sulfur, not sulphur.

R-0.1.2   **Position of locants**

Locants (numerals and/or letters) are placed immediately before the part of the name to which they relate, except in the case of traditional contracted forms (see R-2.5)[4].

Examples:
Hex-2-ene (R-3.1.1) (formerly 2-Hexene)
Cyclohex-2-en-1-ol (R-5.5.1.1) (formerly 2-Cyclohexen-1-ol)
O-Acetyl-*N*-methylhydroxylamine (R-5.4.4)
2-Naphthyl (R-2.5) (a contracted form of Naphthalen-2-yl)
   (not Naphth-2-yl)
2-Pyridyl (R-2.5) (a contracted form of Pyridin-2-yl)
   (not Pyrid-2-yl)

[3] International Union of Pure and Applied Chemistry, Inorganic Chemistry Division, Commission on Nomenclature of Inorganic Chemistry, *Nomenclature of Inorganic Chemistry*, (Recommendations 1990), Blackwell Scientific Publications, Oxford, UK, 1990, 289 pp.
[4] The locant 1 (unity) is often omitted when there is no ambiguity, for example, in 2-chloroethanol. However, in this text, this numeral is always included when another numerical locant appears in the same name.

R-0.1.3   **Punctuation**
               Punctuation in chemical names is frequently of great importance, especially to avoid
               ambiguity.

R-0.1.3.1   *Commas* are used to separate locants that refer to the same part of a name, i.e., locants of
               a series, and to separate letters or letter combinations that denote fusion sites in names of
               fused ring systems.

               Examples:
               1,2-Dichloroethane (R-5.3.1)
               Dibenzo[*a*, *j*]anthracene (R-2.4.1)
               N,*N*-Diethyl-2-furamide (R-5.7.8.1)

R-0.1.3.2   *Full stops* (periods) separate numerical ring size indicators in names constructed accord-
               ing to the von Baeyer system and in certain spiro names.

               Examples:
               Bicyclo[3.2.1]octane (R-2.4.2)
               6-Oxaspiro[4.5]decane (R-2.4.3)

R-0.1.3.3   *Colons* separate related sets of locants; if a higher level of separation is required,
               semicolons are employed.

               Examples:
               1,4,5,8-Tetrahydro-1,4:5,8-dimethanoanthracene (R-2.4.1)
               Benzo[1″,2″:3,4;4″,5″:3′,4′]dicyclobuta[1,2-*b*:1′,2′-*c*′]difuran (R-2.4.1)
               1,1′:2′,1″-Tercyclopropane (R-2.4.4)

R-0.1.3.4   *Hyphens* separate:
               (a) locants from the words or syllables of a name;
               (b) adjacent locants referring to different parts of the name (but preferably parentheses
               should be inserted);
               (c) the two parts of the designation for a primary fusion site in a name for a fused ring
               system;
               (d) a stereodescriptor and the name.

               Examples:
               N-Acetyl-*N*-(2-naphthyl)benzamide (R-5.7.8.2) (preferred to *N*-Acetyl-*N*-2-naphthyl-
                   benzamide)
               2-(3-Pyridyloxy)pyrazine (R-5.5.4)
               Thieno[3,2-*b*]furan (R-2.4.1.1)
               (*E*)-But-2-ene (R-7.1.2)

               After parentheses, a hyphen only appears if the final parenthesis is followed by a locant,
               for example, 3-(bromocarbonyl)-4-(chlorocarbonyl)-2-methylbenzoic acid. Long hy-
               phens are used in certain names (see R-1.2.3.4).

R-0.1.3.5   *Spaces* are a very important type of punctuation for many kinds of names in the English language. If a space is required in a name, it must be used for the name to be unambiguous. On the other hand, the use of spaces where they are not required may be misleading. Spaces are used in names as follows and are not used in any other names.

R-0.1.3.5.1   *Spaces* separate words in most functional class names, the class name being expressed as one word and the remainder of the molecule as one (or more) separate words. Many compounds may be described by such names, such as:

(a) acids and their derivatives, such as salts, esters and anhydrides;

Examples:
Acetic acid (R-9.1, Table 28(a), p. 175)
Potassium sodium succinate (R-5.7.4.1)
Ethyl acetate (R-5.7.4.2)
Phthalic anhydride (R-5.7.7.1)

(b) carbonyl compounds and their derivatives, such as ketones, acetals, hydrazones, and oximes;

Examples:
Ethyl methyl ketone (R-5.6.2.1)
Cyclohexanone ethyl methyl ketal (R-5.6.4.1)
Benzaldehyde oxime (R-5.6.6.1)

(c) halogen and pseudohalogen compounds;

Examples:
tert-Butyl chloride (R-5.3.1)
Methyl cyanide (R-5.7.9.1)
Acetyl chloride (R-5.7.6)

(d) oxygen compounds (alcohols, ethers, peroxides, etc.) and their chalcogen analogues;

Examples:
Ethyl alcohol (R-5.5.1.1)
Ethyl vinyl ether (R-5.5.4)
Ethyl phenyl peroxide (R-5.5.5)
Methyl propyl sulfide (R-5.5.4)
Butyl methyl sulfoxide (R-5.5.7)

R-0.1.3.5.2   *Spaces* separate words in additive names.

Examples:
Styrene oxide (R-1.2.3.3.1)
Trimethylarsane sulfide (R-1.2.3.3.1)

R-0.1.4   **Numerical (multiplicative) prefixes**
               These are derived from Greek and Latin number names and are the principal method for
               describing a multiplicity of identical features of a structure in chemical nomenclature (see
               R-4.1, especially Table 11).

R-0.1.4.1   *The simple numerical prefixes* 'di-', 'tri-', 'tetra-', etc., are of Greek derivation (except for
               'nona-' and 'undeca-', which are derived from Latin) and are used to indicate a
               multiplicity of substituent suffixes, conjunctive components, replacement affixes, simple
               (i.e., unsubstituted) substituent prefixes, and simple (i.e., unsubstituted) functional modifi-
               cation terms provided that there is no ambiguity (see also R-0.1.4.2).

               Examples:
               -diol   ditetradecane-1,14-diyl-
               -dicarboxylic acid   tetra-2-naphthyl-
               tricyclohexyl-   dioxime
               ditridecyl-   dibenzenesulfonate
               -diamido-   ethylenediimino-
               diaza-   diisoxazol-3-yl-
               Benzene-1,3,5-triacetic acid (R-1.2.4.1)

R-0.1.4.2   *The numerical prefixes* 'bis-', 'tris-', 'tetrakis-', etc., which, except for 'bis-' and 'tris-', are
               derived by adding 'kis-' to the simple numerical prefixes (see Table 11), are used to
               indicate a multiplicity of substituted substituent prefixes or functional modification
               terms.

               Examples:
               bis(2-aminoethyl)-
               ethylenebis(oxymethylene)-
               bis(phenylhydrazone)

               Such prefixes are also used when the use of 'di-', 'tri-', etc., is (or could be) ambiguous; this
               usually happens when an analogue of the term being multiplied begins with a simple
               numerical prefix.

               Examples:
               tris(methylene)-
               tris(decyl)-
               bis(ylium) (see R-5.8.2)
               bis(phosphate)
               bis(benzo[*a*]anthracen-1-yl)-
               Benzo[1,2-*c*:3,4-*c'*]bis[1,2,5]oxadiazole

R-0.1.4.3   *The numerical prefixes* 'bi-', 'ter-', 'quater-', etc., are derived from Latin number names
               and are used mainly in ring assembly names.

               Examples:
               Biphenyl (R-2.4.4)
               2,2':6,2'':6,2'''-Quaterpyridine (R-2.4.4)

4

R-0.1.4.4   *The prefix 'mono-'* is usually omitted in chemical names. However, it is used to indicate that only one characteristic group of a parent structure has been modified. The ending '-kis' is not used with 'mono-'.

Examples:
Monoperoxyterephthalic acid (R-5.7.1.3.1)
Phthalic acid monomethyl ester (R-5.7.4.2)

R-0.1.5   **Enclosing marks**
Parentheses (round brackets, curves), square brackets, and braces (curly brackets) are used in chemical nomenclature to set off parts of a name dealing with specific structural features in order to convey the structure of a compound as clearly as possible.

R-0.1.5.1   *Parentheses*
(a) Parentheses are placed around prefixes defining substituted substituents and after the numerical multiplicative prefixes 'bis-', 'tris-', etc.

Examples:

$ClCH_2–SiH_3$
(Chloromethyl)silane

$(HO–CH_2–CH_2–O)_2CH–COOH$
Bis(2-hydroxyethoxy)acetic acid

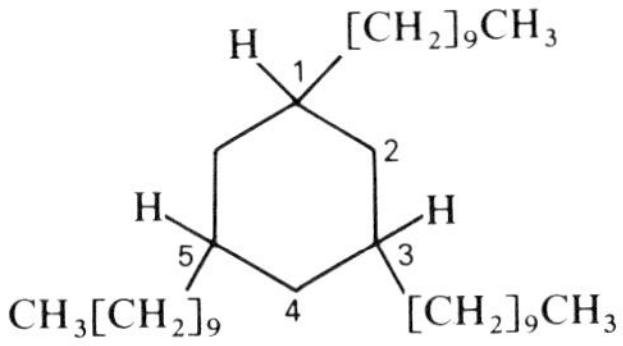

1,3,5-Tris(decyl)cyclohexane **not** 1,3,5-tridecyl-cyclohexane (tridecyl denotes the $CH_3[CH_2]_{12}$-group; tris(decyl) describes three $CH_3[CH_2]_9$-groups)

(b) Parentheses are placed around simple substituent prefixes to separate locants of the same type referring to different structural elements, even though only one may be expressed, and to avoid ambiguity (see also R-0.1.4.2).

Examples:

(2-Naphthyl)phenyldiazene (R-5.3.3.1)

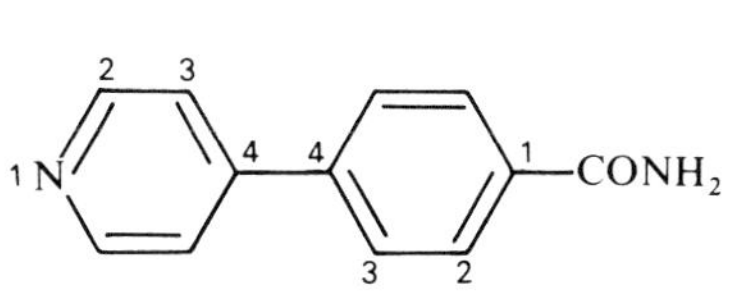

4-(4-Pyridyl)benzamide

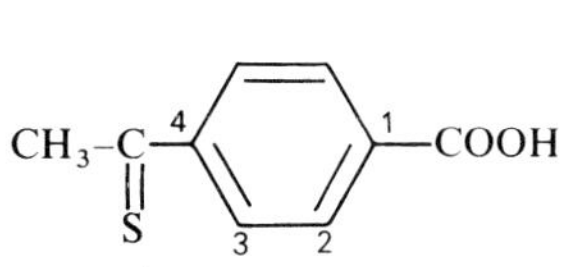

4-(Thioacetyl)benzoic acid

$CH_3–SiH_2Cl$
Chloro(methyl)silane

$HSSC–[CH_2]_4–CSSH$
Hexanebis(dithioic acid)

(c) Parentheses may be used simply to aid clarity.

Examples:
(Thiobenzoic) anhydride (R-5.7.7.1)
$CH_3–CH_2–CH_2–CSSH$ Butane(dithioic) acid

(d) Parentheses are used to isolate the second locant of a double bond when it differs from the first locant by anything other than unity. When locants for a multiple bond differ by only unity, only the first (lower) locant is cited in the name.

Examples:
Hex-2-ene (R-3.1.1)
Bicyclo[6.5.1]tetradec-1(13)-ene (R-3.1.1)

(e) Parentheses are used to enclose 'added hydrogen' and its locant, stereodescriptors, such as *E*, *Z*, *R*, *S*, etc., and in descriptors for isotopically substituted compounds.

Examples:
Pyridin-2(1*H*)-one (R-1.3)
(*E*)-But-2-ene (R-7.1.2)
(^{13}C)Methane (R-8.2.2)

R-0.1.5.2   **Square brackets**
(a) Square brackets enclose descriptors denoting fusion sites in names of fused ring systems, and ring sizes in names constructed according to the von Baeyer system and certain spiro names. They may also enclose ring assembly names when these are followed by a principal group suffix or a suffix describing a parent substituent prefix, and the names of components in certain spiro names.

Examples:
Dibenzo[*b,e*]oxepine (R-2.4.1.1)
Bicyclo[3.2.1]octane (R-2.4.2)
Dispiro[5.1.7.2]heptadecane (R-2.4.3)
[1,2'-Binaphthalene]-2-sulfonic acid (R-5.7.2)
Spiro[cyclopentane-1,1'-indene] (R-2.4.3)

(b) Square brackets enclose locants for structural features of components, such as double bonds in bridges and heteroatoms of component rings in names of fused-ring systems.

Examples:
4a,9a-But[2]enoanthracene (R-2.4.1.2)
4*H*-[1,3]Oxathiolo[5,4-*b*]pyrrole (R-2.4.1.1)

(c) Square brackets enclose substituent prefixes in which parentheses have already been employed.

Example:
[2-(Ethoxycarbonyl)ethyl]trimethylammonium bromide (R-5.7.4.2)

6

(d) Square brackets are also used to enclose descriptors in isotopically labelled compounds.

Example:
[^{13}C]Methane (R-8.3)

(e) Square brackets are employed in formulae to indicate repetition of groups in a chain.

Example:
$CH_3-[CH_2]_{21}-CH_3$
Tricosane (R-2.2.1)

R-0.1.5.3   *Braces* may be used to enclose substituent prefixes in which square brackets have already been used.

Examples:

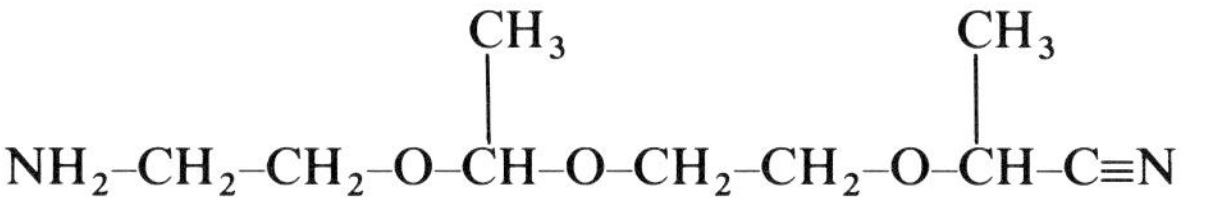

2-{2-[1-(2-Aminoethoxy)ethoxy]ethoxy}propanenitrile

When additional enclosing marks are required, the nesting order is curves, square brackets, braces, then curves, square brackets, braces, etc., i.e. {[()]}, {[({[()]})]}, etc.

Example:

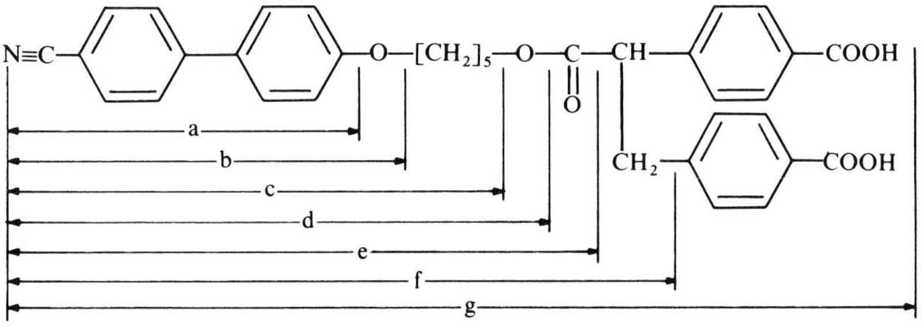

(a)   = 4′-cyanobiphenyl-4-yl
(b)   = (4′-cyanobiphenyl-4-yl)oxy
(c)   = 5-[(4′-cyanobiphenyl-4-yl)oxy]pentyl
(d)   = {5-[(4′-cyanobiphenyl-4-yl)oxy]pentyl}oxy
(e)   = ({5-[(4′-cyanobiphenyl-4-yl)oxy]pentyl}oxy)carbonyl
(f)   = 1-[({5-[(4′-cyanobiphenyl-4-yl)oxy]pentyl}oxy)carbonyl]ethylene
(g)   = 4,4′-{1-[({5-[(4′-cyanobiphenyl-4-yl)oxy]pentyl}oxy)carbonyl]ethylene}-
        dibenzoic acid

R-0.1.6   **Italicization**
Italicizing mainly serves to mark letters which are not involved in the primary stage of alphabetical ordering. In manuscripts, italics are conventionally indicated by underlining.

R-0.1.6.1   *Lower case italic letters* are used in descriptors of fusion sites in names of fused ring systems.

Example:
Thieno[3,2-*b*]furan (R-2.4.1.1)

The lower case italic letters *o*, *m*, *p*, may be used in place of the numerical locants 1,2 (*ortho*), 1,3 (*meta*), and 1,4 (*para*), respectively, for disubstituted benzene derivatives, but the numerals are preferred.

Examples:
o-Dinitrobenzene (R-5.3.2)
p-Aminobenzoic acid (R-5.4.1)

R-0.1.6.2   *Italicized element symbols,* such as *O*-, *N*-, *P*-, *S*-, are locants indicating attachment to these heteroatoms.

Examples:
N-Methylbenzamide (R-5.7.8.1)
O-Ethyl hexanethioate (R-5.7.4.2)

The italic element symbol *H* denotes indicated or added hydrogen.

Examples:
3*H*-Pyrrole (R-1.3)
Phosphinin-2(1*H*)-one (R-1.3)

R-0.1.6.3   *Italic words, syllables and capital roman letters* are used in some structural descriptors and in stereodescriptors.

Examples:
sec (R-5.5.1.1), *tert* (R-5.3.1) (but not 'iso' or 'cyclo')[5], *cis, trans* (R-7.1.1),
r, c, t, (R-7.1.1)
R, S (R-7.2.1); *R** (spoken R-star), *S** (spoken S-star), *rel* (R-7.2.2),
Z, E (R-7.1.2)
abeo (R-1.2.7.1), *retro* (R-1.2.7.2) (but not 'homo', 'nor' or 'seco')

R-0.1.7   **Elision and addition of vowels**

R-0.1.7.1   *Vowels* are systematically elided as follows:

(a) the terminal 'e' in names of parent hydrides when followed by a suffix beginning with 'a', 'i', 'o', 'u', or 'y';

Examples:

Ethanal (R-5.6.1)	Sulfanyl (R-5.8.1.1)
Ethanamine (R-5.4.1)	Methanium (R-5.8.2)
Heptan-2-one (R-5.6.2.1)	Propan-2-ide (R-5.8.3)
Pent-4-en-2-ol (R-4.2.3)	

---

[5] The letters '*s*'-, '*t*'-, and 'i'- used as prefixes with the abbreviation Bu (for butyl) ['i'-also with Pr (for propyl)], can be found in many publications (e.g. *s*-Bu, i-Pr), but they are not recommended for use in names.

(b) in the Hantzsch–Widman system, the final 'a' of an element prefix when followed by a vowel;

Examples:
1,3-Oxazole (R-2.3.3) (**not** 1,3-Oxaazole nor 1,3-Oxaazaole)
1,4-Thiazepine (R-2.3.3) (**not** 1,4-Thiaazepine nor 1,4-Thiaazaepine)

(c) the terminal 'a' in the names of numerical multiplicative affixes when followed by a suffix beginning with 'a' or 'o', or a Hantzsch–Widman prefix or stem beginning with a vowel;

Examples:
Benzenehexol (R-5.5.1.1) (**not** Benzenehexaol)
[1,1′-Binaphthalene]-3,3′,4,4′-tetramine (R-5.4.1) (**not** [1,1′-Binaphthalene]-3,3′,4,4′-tetraamine)
1,3,5,7-Tetraoxocane (R-2.3.3) (**not** 1,3,5,7-Tetraoxaocane)

(d) the terminal 'a' of an element prefix in 'ababa' repeating unit names (see R-2.3.3.2) and the terminal 'o' of a replacement infix when followed by a vowel;

Examples:
Tetrasiloxane (R-5.1.4.2) (**not** Tetrasilaoxane)
P-Phenylphosphonamidimidic acid (R-3.4) (**not** *P*-Phenylphosphonamidoimidic acid)

R-0.1.7.2   There is no elision of vowels in the following cases:

(a) in conjunctive names;

Examples:
Cyclohexaneethanol (R-1.2.4.1)
Cyclopentaneacetic acid (R-1.2.4.1)

(b) from replacement or numerical prefixes in replacement nomenclature;

Example:
2,4,8,10-Tetraoxaundecane (R-2.2.3.1)

(c) from numerical prefixes in multiplying parent compounds;

Example:
Ethylenediaminetetraacetic acid (R-9.1, Table 28(b), p. 176)

(d) from numerical prefixes before substituent prefix names;

Example:
1,3,6,8-Tetraoxo-1,2,3,6,7,8-hexahydropyrene-2-carboxylic acid (R-5.6.2.1)

(e) in composite prefixes;

Examples:
4-(Thioacetyl)benzoic acid (R-5.6.1)
[Oxybis(ethyleneoxy)]diacetic acid (R-4.2.6)

(f) from prefixes designating attached components in fusion nomenclature; for example the terminal 'o' of acenaphtho-, benzo-, naphtho-, perylo-, phenanthro- and the terminal 'a' of anthra-, cyclopropa-, cyclobuta-, etc., are not elided before a vowel as in the 1979 edition of the IUPAC *Nomenclature of Organic Chemistry*[1].

Examples:
Dibenzo[*b,e*]oxepine (R-2.4.1.1) (previously, Dibenz[*b,e*]oxepin)
5*H*-Cyclobuta[*f*]indene (R-2.4.1.1) (previously, 5*H*-Cyclobut[*f*]indene)
Pyrazolo[4′,3′:6,7]oxepino[4,5-*b*]indole (R-2.4.1.1)

R-0.1.7.3   ***Addition of the vowel 'o'.*** For euphonic reasons, the vowel 'o' is sometimes inserted between consonants.

Examples:
Naphthalene-2-sulfonodiimidic acid (R-5.7.2.1)
Ethanesulfonohydroximic acid (R-5.7.2.1)
Diethylphosphinothioic chloride (R-5.7.6)

R-0.1.8   **Order of prefixes**
In general, two types of prefixes are used in naming organic compounds, detachable and nondetachable; each may be subdivided further. For example, two kinds of non-detachable prefixes are those modifying the skeletal structure of a parent, such as 'homo-' and 'nor-', and those indicating replacement of skeletal atoms of a parent hydride, the so-called 'a' prefixes, such as 'aza-' and 'oxa-'. Each class of prefix has its preferred position in front of the name of a parent structure and is ordered as discussed in the following subsections.

R-0.1.8.1   ***Nondetachable prefixes*** that modify the skeletal structure of a parent hydride are cited in alphabetical order immediately preceding the name of the parent hydride. This type of prefix is found commonly in names of natural products (stereoparents) and is discussed more fully in Section F of the IUPAC *Nomenclature of Organic Chemistry*[1]; however, they are occasionally encountered in trivial and semisystematic names, for example, homocubane.

R-0.1.8.2   ***Nondetachable prefixes*** ('a' replacement terms commonly known as 'a' prefixes) that indicate replacement of skeletal atoms of a parent hydride are cited in the order of appearance in R-9.3, immediately preceding nondetachable skeleton-modifying prefixes (see R-0.1.8.1), if any, for example, 7a-oxa-13-aza-7a-homo-18-nor-5α-androstane.

R-0.1.8.3   ***Detachable prefixes*** describing substituents are cited preceding nondetachable prefixes (see R-0.1.8.1 and R-0.1.8.2), if any, and are alphabetized as follows:

10

(a) *Simple prefixes* (i.e., those describing atoms and unsubstituted substituents) are arranged alphabetically; multiplying affixes, if necessary, are then inserted and do not alter the alphabetical order already established.

Examples:

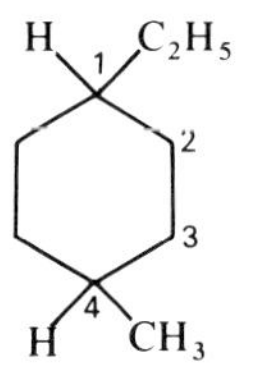

1-Ethyl-4-methylcyclohexane   2,5,8-Trichloro-1,4-dimethylnaphthalene
      ↑   ↑   ↑   ↑

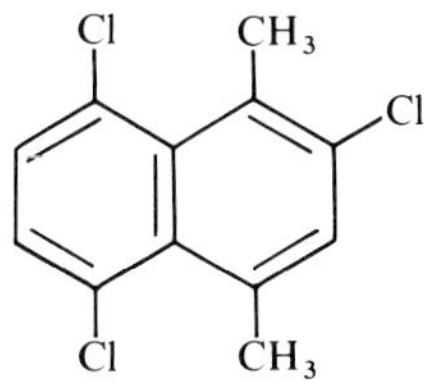

(b) The name of a prefix for a substituted substituent is considered to begin with the first letter of its complete name.

Example:

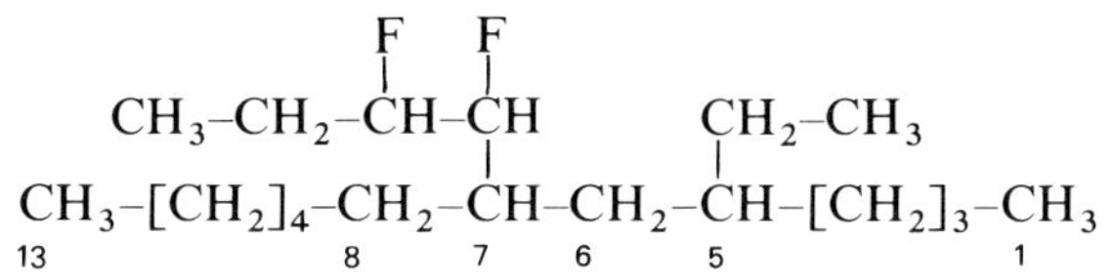

$$CH_3-CH_2-\overset{F}{\underset{|}{CH}}-\overset{F}{\underset{|}{CH}} \qquad CH_2-CH_3$$

$$\underset{13}{CH_3}-[CH_2]_4-\underset{8}{CH_2}-\underset{7}{CH}-\underset{6}{CH_2}-\underset{5}{CH}-[CH_2]_3-\underset{1}{CH_3}$$

7-(1,2-Difluorobutyl)-5-ethyltridecane
      ↑   ↑

(Difluorobutyl as a complete substituent is alphabetized under 'd'.)

(c) When two or more prefixes consist of identical roman letters, priority for citation is given to that group which contains the lowest locant at the first point of difference.

Example:

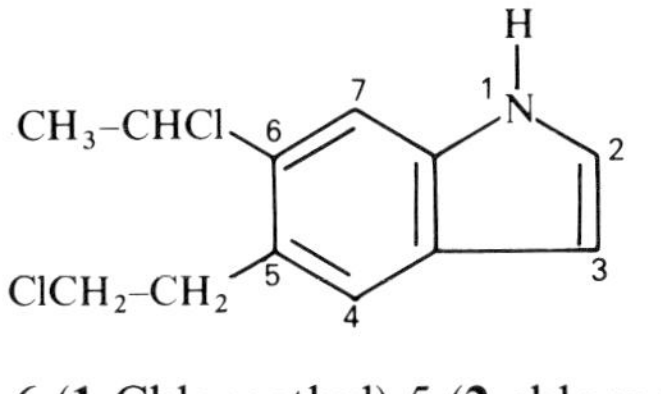

6-(1-Chloroethyl)-5-(2-chloroethyl)-1*H*-indole
    ↑   ↑

(d) *o-* (*ortho*), *m-* (*meta*), and *p-* (*para*) substituents, if otherwise identical, are arranged in that order (the same as their numerical equivalents, 2-, 3-, and 4-, respectively).

11

Example:

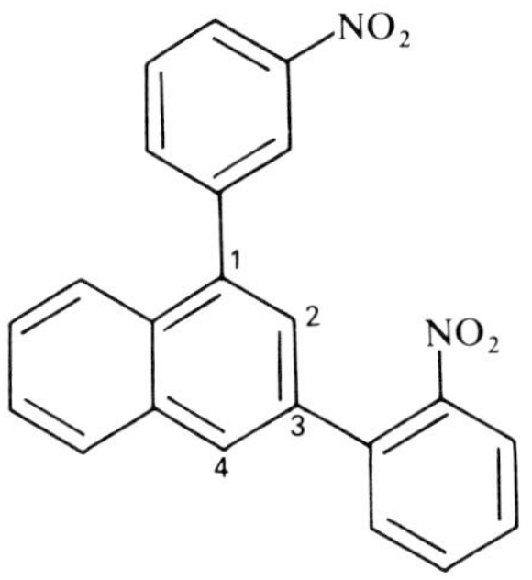

3-(*o*-Nitrophenyl)-1-(*m*-nitrophenyl)naphthalene
↑ ↑

R-0.1.8.4 **Subtractive prefixes** (such as anhydro-, dehydro-, demethyl-) have been used as 'detachable' or 'nondetachable' in previous editions of the IUPAC *Nomenclature of Organic Chemistry*[1], but in these recommendations, they are presented as nondetachable.

Example:

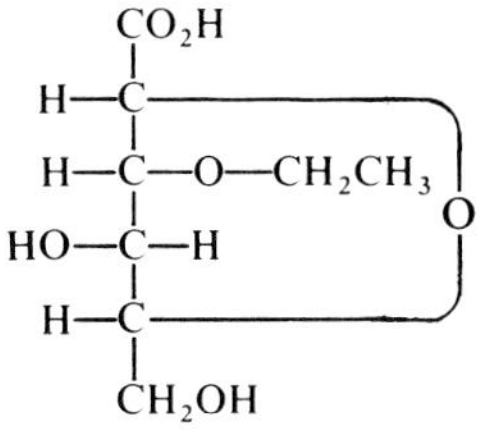

3-*O*-Ethyl-2,5-anhydro-D-gulonic acid (nondetachable)[6]

R-0.1.8.5 **Additive prefixes.** According to the previous edition of the IUPAC *Nomenclature of Organic Chemistry*[1] (C-16.11, p. 108), the additive prefixes 'dihydro-', 'tetrahydro-', etc., could be either detachable and alphabetized among the set of substitutive prefixes, or nondetachable and cited after the substitutive prefixes. In this guide these prefixes are presented as nondetachable[7].

Example:

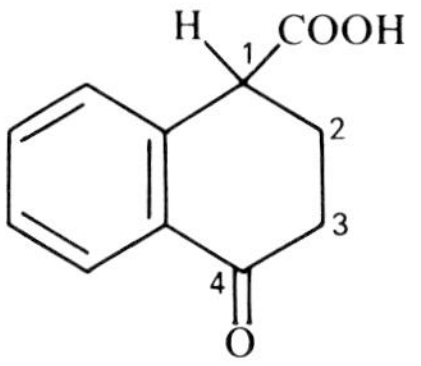

4-Oxo-1,2,3,4-tetrahydronaphthalene-1-carboxylic acid

---

[6] In IUPAC–IUBMB carbohydrate nomenclature, the prefix 'anhydro-' is treated as detachable which gives the following name: 2,5-Anhydro-3-*O*-ethyl-D-gulonic acid; in *Beilstein*, 'anhydro' is nondetachable.
[7] For indexing purposes, Chemical Abstracts Service treats these prefixes as detachable and alphabetizes them along with the substituent prefixes. *Beilstein* also treats them as detachable, but presents them in the name after the substitutive prefixes.

12

R-0.2   GLOSSARY
             Many terms have special meaning in nomenclature. The following are used in this
             document.

R-0.2.1   **Parent structures**

R-0.2.1.1   ***Parent hydride:*** an unbranched acyclic or cyclic structure or an acyclic/cyclic structure
             having a semisystematic or trivial name to which *only* hydrogen atoms are attached.

             Examples:
             Methane (R-2.1)
             Cyclohexane (R-2.3.1.1)
             Styrene (R-9.1, Table 19(a), p. 163)
             Pyridine (R-9.1, Table 23, p. 169)

R-0.2.1.2   ***Functional parent:*** a structure the name of which implies the presence of one or more
             characteristic groups and which has one or more hydrogen atoms attached to at least
             one of its skeletal atoms or one of its characteristic groups, or in which at least one of its
             characteristic groups can form at least one kind of functional modification.

             Examples:
             Acetic acid (R-9.1, Table 28(a), p. 175)
             Aniline (R-9.1, Table 29(a), p. 177)
             Phosphonic acid (R-3.3)

             Note: A parent hydride bearing a characteristic group denoted by a suffix, for example,
             cyclohexanol, is not considered to be a functional parent, but may be described as a
             'functionalized parent hydride'.

R-0.2.2   **Groups**

R-0.2.2.1   ***Substituent atom or group*** (formerly 'radical'[8]): an atom or group that replaces one or
             more hydrogen atoms attached to a parent structure or characteristic group except for
             hydrogen atoms attached to a chalcogen atom.

R-0.2.2.2   ***Characteristic group:*** a single heteroatom, for example, $-Cl$ and $=O$; a heteroatom
             bearing one or more hydrogen atoms or other heteroatoms, for example, $-NH_2$, $-OH$,
             $-SO_3H$, $-PO_3H_2$, and $-IO_2$; or a heteroatomic group attached to or containing a single
             carbon atom, for example, $-CHO$, $-C\equiv N$, $-COOH$, and $-NCO$, attached to a parent
             hydride. The most common of these groups are listed in Tables 5 and 9.

R-0.2.2.3   ***Principal group:*** the characteristic group chosen for citation at the end of a name by a
             suffix or a class name.

---

[8] The term 'radical' is now restricted to species containing unpaired electrons.

R-0.2.3   **Names**

R-0.2.3.1   *Trivial name:* a name having no part used in a systematic sense.

               Example:
               Urea (R-9.1, Table 31(a), p. 179)

R-0.2.3.2   ***Semisystematic name or semitrivial name***: a name in which at least one part is used in a
               systematic sense.

               Examples:
               Glycerol (ol) (R-9.1, Table 26(a), p. 173)
               Acetone (one) (R-9.1, Table 27(a), p. 174)
               Styrene (ene) (R-9.1, Table 19(a), p. 162)
               5α-Cholestane (ane) (R-2.4.6)

R-0.2.3.3   ***IUPAC name:*** a name formed according to the procedures described in Sections A, B, C,
               D, E, F, and H of the 1979 edition of the IUPAC *Nomenclature of Organic Chemistry*[1] as
               modified by these recommendations (see also R-1). There are several kinds of IUPAC
               names according to the type of nomenclature operation involved (see R-1.2).

R-0.2.3.3.1   *Fusion name:* a composite name for a polycyclic parent structure having the maximum
               number of noncumulative double bonds and at least one *ortho* fusion. Name formation
               involves the dissection of the structure into contiguous components having recognized
               trivial or semisystematic names, one of which is selected as the 'base component'.
               Attachment of the other component(s) is described by prefixes (see R-2.4.1).

               Examples:
               Dibenzo[*b,e*]oxepine (R-2.4.1)
               Thieno[3,2-*b*]furan (R-2.4.1)

R-0.2.3.3.2   *Hantzsch–Widman name:* a name for a heteromonocyclic parent hydride having no more
               than 10 ring members formed by the citation of 'a' prefixes denoting the heteroatoms
               followed by an ending (the 'stem') defining the size of the ring (see R-2.3.3).

               1,3-Dithiolane (R-2.3.3)
               1,4-Thiazepine (R-2.3.3)

R-0.2.3.3.3   *Functional class name:* a name that expresses a characteristic group as a class term
               written as a separate word following the name of a parent structure or a name derived
               from a parent structure. In the latter case, when the derived name is that for a substituent
               group (formerly called a 'radical'), the method has been called 'radicofunctional nomen-
               clature' (see R-1.2.3.3.2).

               Examples:
               Methyl iodide (R-5.3.1)   Trimethylarsane sulfide (R-1.2.3.3.1)
               Ethyl alcohol (R-5.5.1.1)   Propanal hydrazone (R-5.6.6.2)
               Ethyl methyl ketone (R-5.6.2.1)   Ethyl acetate (R-5.7.4.2)
               Acetyl chloride (R-5.7.6)

R-0.2.3.3.4   *Radicofunctional name:* see under 'Functional class name'.

R-0.2.3.3.5   *Replacement name:* a name in which the replacement of an atom or a group of a parent structure by another atom or group is indicated by affixes attached to or inserted into the name of the parent structure. There are two main types of replacement names.

(a) *Skeletal replacement name:* a name in which the replacement of skeletal atoms and their associated hydrogen atoms of a parent hydride by other atoms with the appropriate number of hydrogen atoms is indicated by nondetachable prefixes. When carbon atoms are replaced by heteroatoms (see R-1.2.2), this method has been called 'a' nomenclature, since the prefixes used end in 'a'.

Examples:
Silacyclohexane (R-1.2.2.1)
2,7,9-Triazaphenanthrene (R-1.2.2.1)

Certain names in which the prefix 'thio-', seleno-', or 'telluro-' indicates replacement of a skeletal oxygen atom by a sulfur, selenium, or tellurium atom, respectively, are also skeletal replacement names.

Example:
2*H*-Thiopyran (R-1.2.2.2)

(b) *Functional replacement name:* a name containing prefixes or infixes which indicate the replacement of an oxygen atom or hydroxy group of a characteristic group, functional parent, or class name by other atoms or groups.

Examples:
Selenobenzoic acid (R-1.2.2.2)
Hexane(dithioic) acid (R-1.2.2.2)
Methyl *P,P*-dimethylphosphinimidate (R-1.2.2.2)

R-0.2.3.3.6   *Substitutive name:* a name which indicates the exchange of one or more hydrogen atoms attached to a skeletal atom of a parent structure or to an atom of a characteristic group for another atom or group, which may be expressed by a suffix or by prefixes (see R-1.2.1).

Examples:
9,10-Diphenylanthracene (R-5.1.1)
Butane-1,4-diol (R-5.5.1.1)
Ethylphosphonic acid (R-5.7.3.1)
N-Methylbenzamide (R-5.7.8.1)

R-0.2.3.3.7   *Conjunctive name:* a name for assemblies of functionalized acyclic parent hydrides and cyclic systems implying the loss of an appropriate number of hydrogen atoms from each (see R-1.2.4).

Examples:
Cyclohexaneethanol (R-1.2.4.1)
Benzene-1,3,5-triacetic acid (R-1.2.4.1)

R-0.2.3.3.8   *Additive name:* a name that describes:

(a) the formal assembly of names for the components of a compound without loss of atoms or groups of atoms from any component, as in the following:

Functional class name:
Styrene oxide (R-1.2.3.3.1)
Methyl iodide (R-5.3.1)
Ethyl methyl ether (R-5.5.4)

Ring assembly names:
Biphenyl (R-2.4.4)
2,2'-Bipyridyl (R-2.4.4)

Acyclic 'assembly' name:
Biacetyl (R-9.1, Table 27(a), p. 174)

Salt names:
Calcium diacetate (R-5.7.4.1)
[2-(Ethoxycarbonyl)ethyl]trimethylammonium bromide (R-5.7.5.2)

Compound substitutive or multiplicative prefixes[9]:
Pentyloxy (R-5.5.2)
Methylenedioxy (R-5.5.2)

(b) the addition or attachment of atoms or groups of atoms, as illustrated by the following:

Skeleton modifying name:
4a-Homo-5α-pregnane (R-1.2.3.1)

Prefixes or suffixes that increase the number of substitutable hydrogen atoms of a parent hydride, such as hydro- and -ium:

Examples:
2,3,4,5-Tetrahydroazocine (R-3.1.2)
Pyridinium (R-1.2.3.2)

R-0.2.3.3.9   *Subtractive name:* a name for a modified parent structure in which prefixes and/or suffixes indicate the removal of atoms or groups and, where required, replacement by an appropriate number of hydrogen atoms (see R-1.2.5).
Prefixes: de-, anhydro-, nor-.
Suffixes that decrease the number of substitutable hydrogen atoms of a parent structure: -ene, -yne, -ylium, -yl, -ide.

---

[9] Names such as 'methylthio' used previously in the **IUPAC** *Nomenclature of Organic Chemistry*[1] are additive names, but are not encouraged in these recommendations.

Examples:
Demethylmorphine (R-1.2.5.1)
3-Norlabdane (R-1.2.5.1)
Hex-2-ene (R-3.1.1)
Methyl (R-5.8.1.1)

R-0.2.3.3.10   *Multiplicative name:* a name that expresses multiple occurrence of identical parent structures, two or more of which are connected by a symmetrical structure expressible by means of a multivalent simple or composite prefix (see R-1.2.8).

Examples:
4,4′-Peroxydibenzoic acid (R-5.5.5)
[Oxybis(ethyleneoxy)]diacetic acid (R-4.2.6)

R-0.2.4   **Other terms used in these recommendations**

R-0.2.4.1   ***Seniority, senior:*** terms used in reference to priority in a prescribed hierarchical order, a senior feature being preferred.

R-0.2.4.2   ***Lowest set of locants:*** the set of locants, which when compared term by term with other locant sets, each in order of increasing magnitude, has the lowest term at the first point of difference; for example, the locant set 2,3,6,8 is lower than 3,4,6,8 or 2,4,5,7.
        Primed locants are placed immediately after the corresponding unprimed locants in a set arranged in ascending order; locants consisting of a numeral and a lower-case letter are placed immediately after the corresponding numeric locant.

Examples:
2 is lower than 2′
3 is lower than 3a
8a is lower than 8b
4′ is lower than 4a
4′a is lower than 4a′

Italic capital and lower-case letter locants are lower than Greek letter locants which, in turn, are lower than numerals.

Example:
N,α,1,2 is lower than 1,2,4,6

R-0.2.4.3   ***Bonding number:*** see R-1.1.

R-1   General Principles of Organic Nomenclature

R-1.0   INTRODUCTION

Systematic naming of an organic compound generally requires the identification and naming of a parent structure. This name may then be modified by prefixes, infixes, and, in the case of a parent hydride, suffixes, which convey precisely the structural changes required to generate the actual compound from the parent structure.

   Most commonly, a *parent structure* is a parent hydride, i.e., a structure containing, besides hydrogen, either a single atom of an element, for example, phosphane; a number of atoms (alike or different) linked together to form an unbranched chain, for example, pentane and disiloxane; or a monocyclic or polycyclic ring system, for example, cyclohexane, pyridine, naphthalene, and quinoline. It is sometimes convenient to employ parent hydrides of more complex structure such as ring assemblies or ring/chain systems, for example, biphenyl, styrene, ferrocene, and cyclophanes, and to include structures with implied stereochemistry (stereoparents), for example, 5α-cholestane[10]. Rules for naming parent hydrides are given in R-2; in addition, a special class of parent structures termed functional parents, for example, phosphinic acid, is considered in R-3.3. Examples of parent structures are shown below:

PH_3 $CH_3-CH_2-CH_2-CH_2-CH_3$ $SiH_3-O-SiH_3$
Phosphane   Pentane   Disiloxane   Cyclohexane

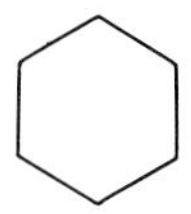

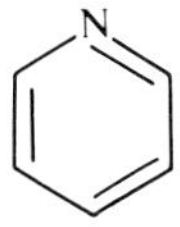

Pyridine 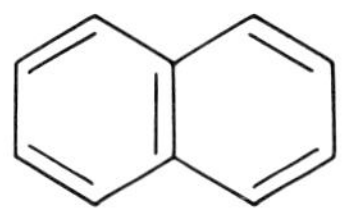   Naphthalene 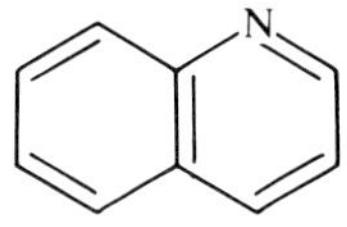   Quinoline

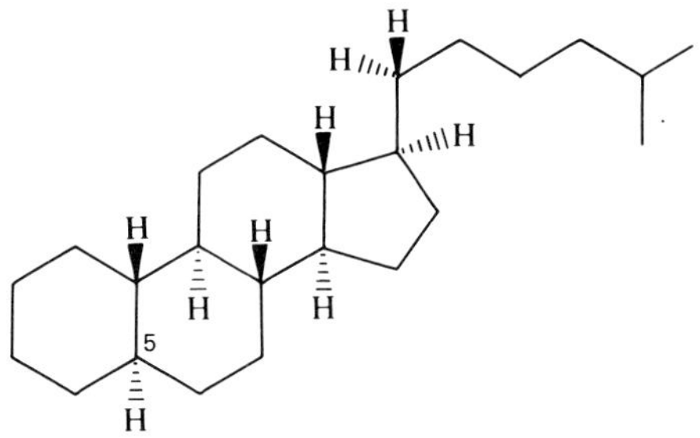

5α-Cholestane[11,12]

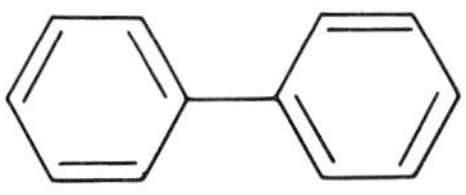

Biphenyl 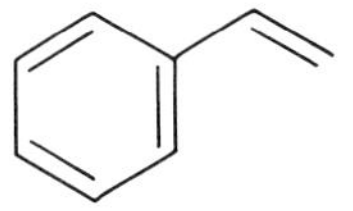   Styrene H_2PO_2H CH_3-COOH
                                                                        Phosphinic acid   Acetic acid

---

[10] The Greek letters α and β are used as stereodescriptors in steroid names (see R-7.1.0).

[11] By convention, a broken wedged line denotes a bond projecting behind the plane of the paper and a solid wedged line denotes a bond projecting in front of the plane of the paper; a normal line denotes a bond lying in the plane of the paper.

[12] For the sake of simplicity, stereochemical configuration is specified only in selected examples in Sections R-1 to R-6.

In order to generate the parent structure from a molecule to be named, various formal operations must be carried out. For example, in naming the structure below,

$$ClCH_2\text{--}CH_2\text{--}CH_2\overset{\overset{\text{O}}{\|}}{\text{--}C}\text{--}CH_3$$
$$54321$$

the parent hydride 'pentane' is formally derived by replacing the oxygen and chlorine atoms by the appropriate number of hydrogen atoms. For constructing a name, this

Table 1 Examples of nomenclature operations

$$CH_3\text{--}CH_2\text{--}O\text{--}CH_2\text{--}CH_2\text{--}CH_3 \qquad CH_3\text{--}CH_2\text{--}CH_2\overset{\overset{\text{O}}{\|}}{\text{--}C}\text{--}CH_2Cl$$

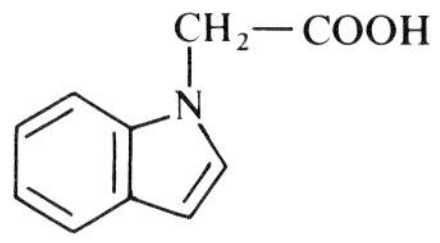

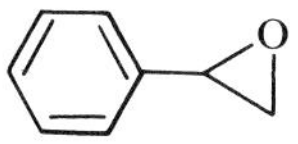

1   2   3

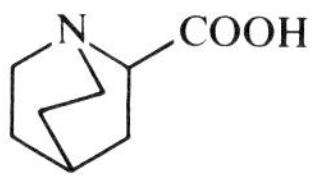

$P(OCH_3)_3$

4   5   6   7 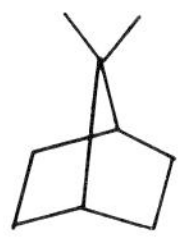   8

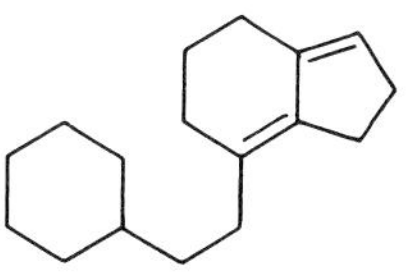

Structure	Parent structure (Class Name)	Operation	Name	Reference
1	Propane (ether)	substitutive functional class[a]	1-Ethoxypropane Ethyl propyl ether	R-1.2.1 R-1.2.3.3.2
2	Pentane (ketone)	substitutive functional class[a]	1-Chloropentan-2-one Chloromethyl propyl ketone	R-1.2.1 R-1.2.3.3.2
3	Acetic acid	substitutive	1H-Indol-1-ylacetic acid	R-1.2.1
	Indole and } Acetic acid }	conjunctive	1H-Indole-1-acetic acid	R-1.2.4.1
4	Styrene (oxide)	additive	Styrene oxide	R-1.2.3.3
	Oxirane	substitutive	2-Phenyloxirane	R-1.2.1
5	Quinuclidine	substitutive	Quinuclidine-2-carboxylic acid	R-1.2.1
	Bicyclooctane	{ replacement	1-Azabicyclo[2.2.2] octane-2-carboxylic acid	R-1.2.2.1
		{ substitutive		R-1.2.1
6	(phosphite)	functional class	Trimethyl phosphite	R.1.2.3.3.2
	Phosphane	substitutive	Trimethoxyphosphane	R-1.2.1
		coordination (additive)	Trimethoxophosphorus	R-1.2.3.1
7	Gonane	{ ring cleavage	9,10-Secogona-8(14),13(17)-diene	R-1.2.6.2
		{ subtractive		R-1.2.5.2
	Indene	{ substitutive	7-(2-Cyclohexylethyl)-2,4,5,6-tetrahydro-1H-indene	R-1.2.1
		{ additive		R-1.2.3.1
8	Bornane	subtractive	10-Norbornane	R-1.2.5.1
	Bicycloheptane	substitutive	7,7-Dimethylbicyclo-[2.2.1]heptane	R-1.2.1

[a] Formerly 'radicofunctional names'.

formal operation is reversed; the prefix 'chloro-' and the suffix '-one' indicating substitution of hydrogen atoms of pentane are attached to the parent hydride name, giving the name *5-chloropentan-2-one*. Prefixes and suffixes can represent a number of different types of formal operations on the parent structure. These are defined in R-1.2. Frequently, the prefix or suffix denotes the attachment of a characteristic group, for example, 'oxo-' or '-one' for =O; lists of such affixes are given in R-3.2. A prefix may describe a group which is derived from a parent hydride, for example, pentan-1-yl or pentyl for $CH_3-CH_2-CH_2-CH_2-CH_2-$ (from pentane); such prefixes are described in R-2.5.

The substitutive operation, described in R-1.2.1, is the operation used most extensively in organic nomenclature. Indeed, the comprehensive nomenclature system based largely on the application of this operation to parent structures is, for convenience, termed 'substitutive nomenclature', although this system also involves many of the other types of operations described in R-1.2. Examples of this and other nomenclature operations are shown in Table 1.

In constructing the names described in R-1.2.3.3.2 (formerly called 'radicofunctional names'), the characteristic group of the compound is expressed as a functional class name, and is usually cited as a separate word rather than as a suffix. In these recommendations, however, names obtained by a substitutive operation are preferred.

The replacement operation can be used for naming organic compounds in which skeletal atoms of a parent structure are replaced by other skeletal atoms, or in which oxygen atoms and/or hydroxy groups of characteristic groups are replaced by other atoms or groups.

It is very important to recognize that, in general, the rules of organic nomenclature are written in terms of classical valence bonding and do not imply electronic configurations of any kind.

Examples of naming structures in several ways are shown in Table 1.

Full details of the way in which parent names may be combined with appropriate prefixes and suffixes are given in R-4 (Name Construction); rules for selection of a unique systematic name, if required, will be described in a separate document. Methods for the specification of stereochemistry are given in R-7 and those for denoting isotopic modification are described in R-8.

R-1.1   BONDING NUMBER

The concept of a standard valence state is fundamental to organic nomenclature. Since most organic names are derived by formal exchange of atoms or groups for hydrogen atoms of a parent structure, it is necessary to know exactly how many hydrogen atoms are implied by the name of the parent structure. For example, does the name phosphane refer to PH_3 or to PH_5? This is a problem only when an element can occur in more than one valence state; in such cases the 'standard' state is normally not specified, but any other valence state is noted by citation of the appropriate *bonding number*. A more detailed treatment of bonding number, including selection rules, is given in separate publications[13,14].

[13] International Union of Pure and Applied Chemistry. Organic Chemistry Division. Commission on Nomenclature of Organic Chemistry, 'Treatment of Variable Valence in Organic Nomenclature (Lambda-Convention) (Recommendations 1983)', *Pure Appl. Chem.*, **56**, 769–778 (1984).

[14] International Union of Pure and Applied Chemistry. Organic Chemistry Division. Commission on Nomenclature of Organic Chemistry, 'Nomenclature for Cyclic Organic Compounds with Contiguous Formal Double Bonds (δ-Convention) (Recommendations 1988)', *Pure Appl. Chem.*, **60**, 1395–1401 (1988).

R-1.1.1 **Definition**

The bonding number 'n' of a skeletal atom is the sum of the total number of bonding equivalents (valence bonds) of that skeletal atom to adjacent skeletal atoms in a parent hydride, if any, and the number of attached hydrogen atoms, if any.

Examples:

SH_2	for S, $n = 2$
SH_6	for S, $n = 6$
$(C_6H_5)_3PH_2$	for P, $n = 5$

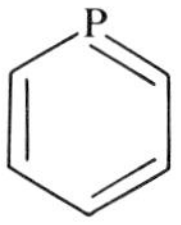

for P, $n = 3$

R-1.1.2 **Standard bonding numbers**

The bonding number of a skeletal atom is *standard* when it has the value given in the following table:

Standard bonding number (n)	Element				
3	B				
4	C	Si	Ge	Sn	Pb
3	N	P	As	Sb	Bi
2	O	S	Se	Te	Po
1	F	Cl	Br	I	At

R-1.1.3 **Nonstandard bonding numbers**

A nonstandard bonding number of a *neutral* skeletal atom in a parent hydride is indicated by the symbol λ^n, cited in conjunction with an appropriate locant.

Examples:

$CH_3–SH_5$
Methyl-λ^6-sulfane

$(C_6H_5)_3PH_2$
Triphenyl-λ^5-phosphane

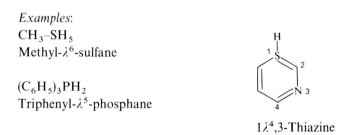

$1\lambda^4,3$-Thiazine

R-1.1.4 **Contiguous formal double bonds**

The presence of contiguous formal double bonds at a skeletal atom in a cyclic parent hydride whose name normally implies the maximum number of noncumulative double bonds is described by the symbol δ^c, where c is an arabic numeral representing the number of double bonds. The δ^c symbol is cited immediately after an expressed locant for the skeletal atom in the name of the parent hydride and follows the λ^n symbol, if present.

Examples:

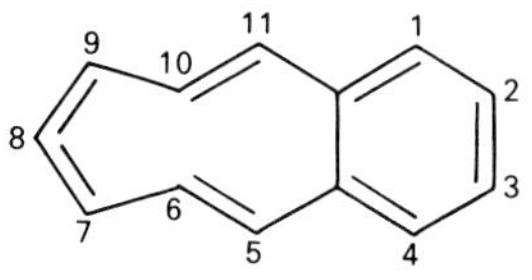

$8\delta^2$-Benzo[9]annulene 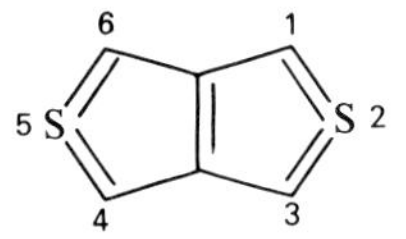 $2\lambda^4\delta^2,5\lambda^4\delta^2$-Thieno[3,4-c]thiophene

R-1.2   NOMENCLATURE OPERATIONS

The operations described in this section all involve structural modification, and are classified first according to the type of modification, for example, 'replacement'; and then according to the way in which the modification is expressed, for example, 'by use of replacement infixes'. The structures to which the various modifications are applied can be regarded as parent structures, and the modifications are expressed by affixes, i.e., infixes, prefixes, and suffixes. General principles involved in identifying a parent structure are described briefly in Section R-4 (Name Construction) and will be described more fully in a separate publication (Name Selection, in preparation).

Examples:

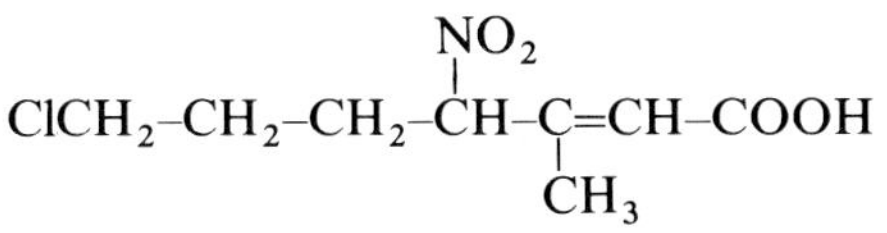

Parent:

$CH_3-CH_2-CH_2-CH_2-CH_2-CH_2-CH_3$
Heptane

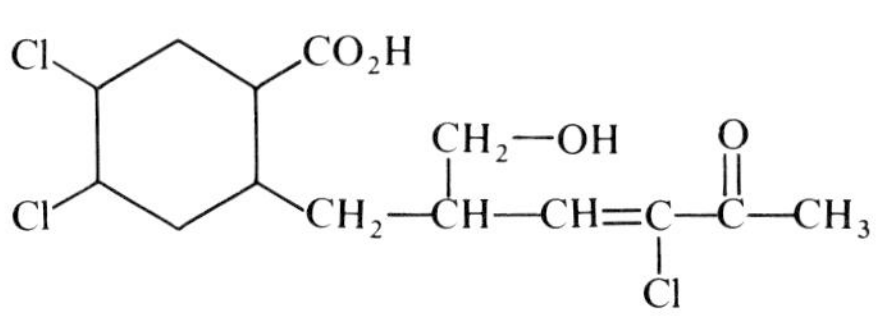

Parent:
Cyclohexane

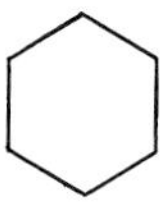

(Cyclohexane is chosen as parent because it bears the principal characteristic group –COOH, see R-4).

R-1.2.1   Substitutive operation

The substitutive operation involves the exchange of one or more hydrogen atoms for another atom or group. This process is expressed by a prefix or suffix denoting the atom or group being introduced (see R-3.2 and R-4 for lists of prefixes and suffixes).

Examples:

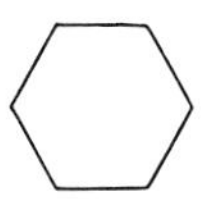

Cyclohexane

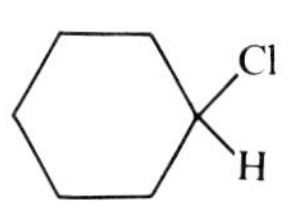

Chlorocyclohexane
(substitutive prefix = chloro)

CH_3-CH_3
Ethane

CH_3-CH_2-SH
Ethanethiol
(substitutive suffix = thiol)

R-1.2.2   **Replacement operation**
               The replacement operation involves the exchange of one group of atoms or a single non-hydrogen atom for another. This can be expressed in several ways as follows:

R-1.2.2.1   *By use of 'a' prefixes representing the elements(s) being introduced,* usually signifying replacement of carbon (see R-9.3).

               Examples:

Cyclohexane   Silacyclohexane
                                        (replacement prefix = sila)

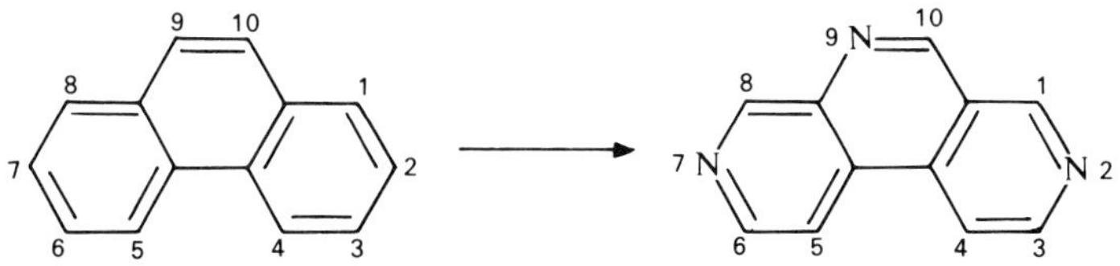

Phenanthrene   2,7,9-Triazaphenanthrene
                                       (replacement prefix = aza)

               Note: Except for bismuth, a replacement term for a cationic atom of the nitrogen, chalcogen or halogen families of elements is denoted by replacing the final 'a' of the corresponding replacement term for the neutral heteroatom by '-onia'; the cationic replacement term for bismuth, the 'a' term for which is 'bisma', is 'bismuthonia' (see R-5.8.2).

R-1.2.2.2   *By use of prefixes or infixes signifying replacement of oxygen atoms or oxygen-containing groups* (see R-3.4). These affixes represent the group(s) being introduced.

               Examples:

$(CH_3)_2P(O)(OCH_3)$ $\longrightarrow$ $(CH_3)_2P(NH)(OCH_3)$
Methyl dimethylphosphinate   Methyl *P,P*-dimethylphosphinimidate
                                                            (replacement infix = imid(o))

$C_6H_5P(O)(OH)_2$ $\longrightarrow$ $C_6H_5P(N)(OH)$

Phenylphosphonic acid   Phenylphosphononitridic acid
                                                            (replacement infix = nitrid(o))

               The chalcogen affixes 'thio-', 'seleno-', and 'telluro-' indicate replacement of an oxygen atom by another chalcogen atom[15].

---

[15] The distinction between these replacement terms and the corresponding 'a' replacement terms, thia, selena, and tellura, respectively, which indicate replacement of *skeletal carbon atoms*, must be noted.

Examples:

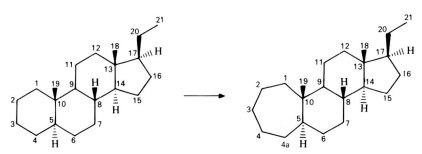

$$\underset{\text{Benzoic acid}}{C_6H_5-\overset{\overset{\displaystyle O}{\|}}{C}-OH}$$

$$\longrightarrow$$

$$\underset{\text{Selenobenzoic acid}}{C_6H_5-C\left\{\begin{matrix}Se\\O\end{matrix}\right\}H}$$

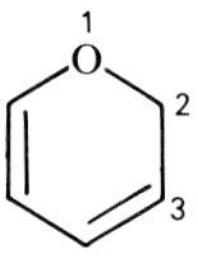

2*H*-Pyran 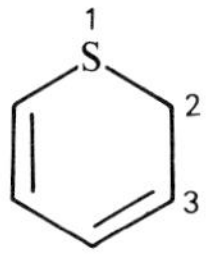   2*H*-Thiopyran

$$\underset{\text{Hexanoic acid}}{CH_3-[CH_2]_4-COOH}$$ $$\longrightarrow$$ $$\underset{\text{Hexane(dithioic) acid}}{CH_3-[CH_2]_4-CSSH}$$

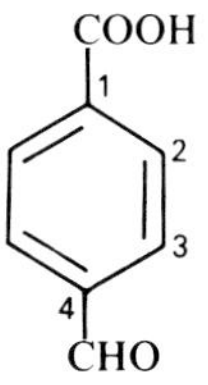

4-Formylbenzoic acid 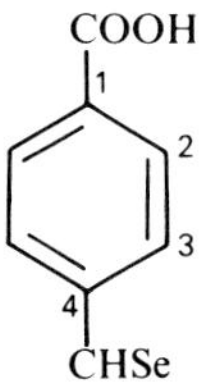   4-(Selenoformyl)benzoic acid

R-1.2.3 Additive operation

The additive operation involves the formal assembly of a structure from its component parts without loss of any atoms or groups. This can be expressed in various ways as follows:

R-1.2.3.1 *By use of an additive prefix*

Examples:

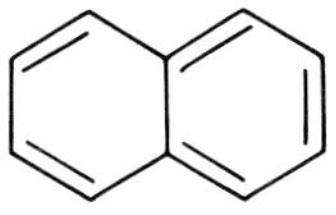

Naphthalene

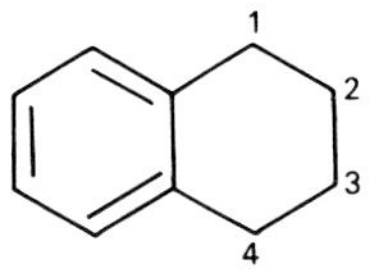

1,2,3,4-Tetrahydronaphthalene
(hydro = addition of one H atom)

5α-Pregnane

4a-Homo-5α-pregnane
(homo = addition of a CH$_2$ (methylene) group, in this case to expand a ring)

Note: Coordination nomenclature, used extensively in nomenclature for inorganic compounds, is an additive operation.

Examples:

Pt^{2+} + $2Cl^-$
Platinum   Dichloro
+ 2 $(C_2H_5)_3P$
Bis(triethylphosphine)
$\longrightarrow$ $[PtCl_2[P(C_2H_5)_3]_2]$

Dichlorobis(triethylphosphine)platinum
(chloro = addition of one Cl atom;
triethylphosphine = addition of one
$(C_2H_5)_3P$ ligand group)

R-1.2.3.2   **By use of an additive suffix**

Examples:

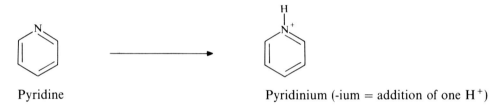

Pyridine   Pyridinium (-ium = addition of one H^+)

R-1.2.3.3   **By use of a separate word**

R-1.2.3.3.1   *With the name of a neutral parent structure*

Examples:

$(CH_3)_3As$ $\longrightarrow$ $(CH_3)_3AsS$
Trimethylarsane   Trimethylarsane sulfide

$C_6H_5CH=CH_2$ $\longrightarrow$ $C_6H_5-\overset{\displaystyle O}{\overset{\displaystyle /\backslash}{CH-CH_2}}$
Styrene   Styrene oxide

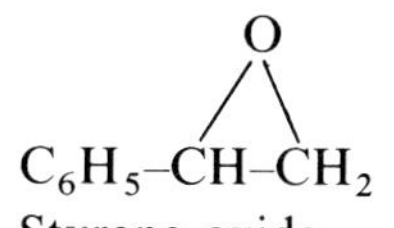

R-1.2.3.3.2   *With one or more substituent prefix ('radical') name(s) (formerly called radicofunctional nomenclature)*[16]. Here the separate word is a class or subclass name representing the characteristic group or the kind of characteristic group to which the substituents ('radicals') are linked.

---

[16] It is convenient to classify the main operation involved in functional class nomenclature as an additive one, as is done here. However, it is also possible (and probably more relevant from a historical point of view) to regard the process as one of specifying the 'radicals' present in compounds for which a class name is given. For instance, the name methyl alcohol (for CH_3-OH) consists of the name methyl for the 'radical' CH_3- and the class name alcohol (for R–OH).

Examples:

$$CH_3- \ + -OH \quad \longrightarrow \quad CH_3-OH$$
Methyl   Alcohol   Methyl alcohol

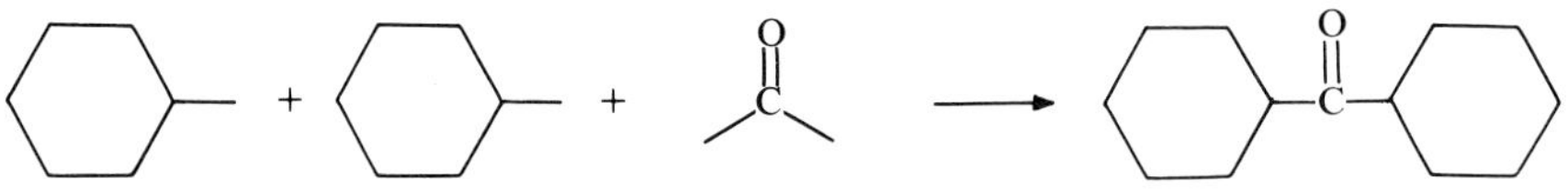

Cyclohexyl   Cyclohexyl   Ketone   Dicyclohexyl ketone

$$CH_3- \ + C_6H_5- \ + -O- \quad \longrightarrow \quad CH_3-O-C_6H_5$$
Methyl   Phenyl   Ether   Methyl phenyl ether

$$C_6H_5-CH_2- \ + \ -C{\equiv}N \quad \longrightarrow \quad C_6H_5-CH_2-C{\equiv}N$$
Benzyl   Cyanide   Benzyl cyanide

R-1.2.3.4   *By connecting the names of the components of an addition compound with a dash (long hyphen)*

Example:

$$CO \qquad + \qquad BH_3 \quad \longrightarrow \quad CO \cdot BH_3$$
Carbon monoxide   Borane   Carbon monoxide—borane

R-1.2.3.5   *By juxtaposition or multiplication of substituent prefix terms*

Examples:

$$CH_3-CH_2-CH_2-CH_2-CH_2- \ + -O- \longrightarrow CH_3-CH_2-CH_2-CH_2-CH_2-O-$$
Pentyl   Oxy   Pentyloxy

$$Cl- \qquad + \quad -CO- \quad \longrightarrow \quad Cl-CO-$$
Chloro   Carbonyl   Chlorocarbonyl

$$-NH- \ + \ -CH_2-CH_2- \ + -NH- \longrightarrow -NH-CH_2-CH_2-NH-$$
Imino   Ethylene   Imino   Ethylenediimino

$$C_6H_5- \qquad + \qquad C_6H_5- \quad \longrightarrow \quad C_6H_5-C_6H_5$$
Phenyl   Phenyl   Biphenyl

R-1.2.4   **Conjunctive operation**

The conjunctive operation involves the formal construction of the name of a compound from those of its components with abstraction of the same number of hydrogen atoms from each component at each site of conjunction. This operation may be expressed:

R-1.2.4.1   *By juxtaposition of component names.* This method is most commonly used when the two components to be joined are: (a) a ring or a ring system; and (b) a carbon chain (or chains) substituted by a principal characteristic group. In this method, both the principal characteristic group and the ring or ring system must terminate the chain; the rest of the structure attached to the chain, if any, is described by substituent prefixes, the location of which is indicated by Greek letter locants, α, β, etc., (α designating the atom next to the principal characteristic group).

Examples:

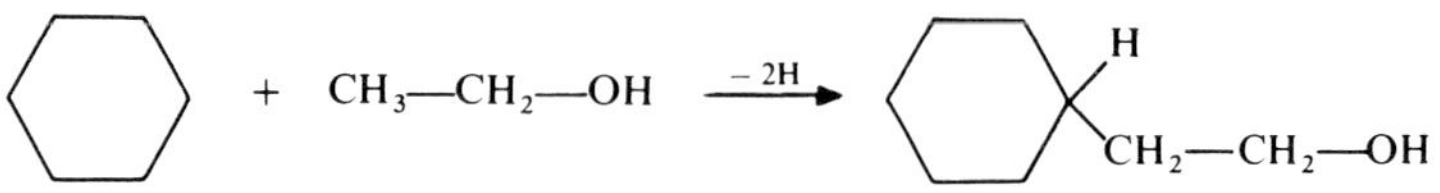

Cyclohexane   Ethanol   Cyclohexaneethanol

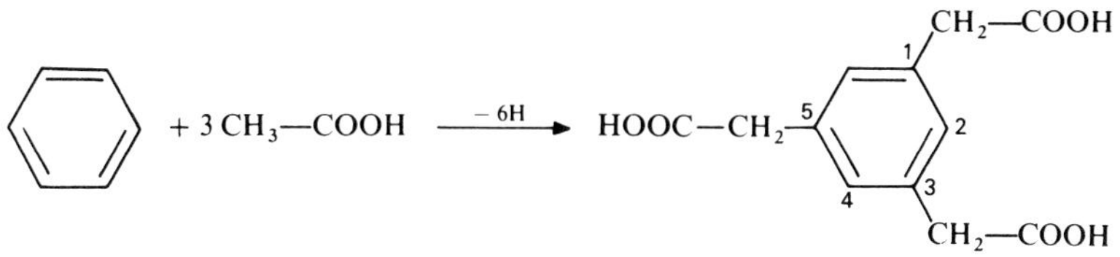

Benzene   Acetic acid   Benzene-1,3,5-triacetic acid

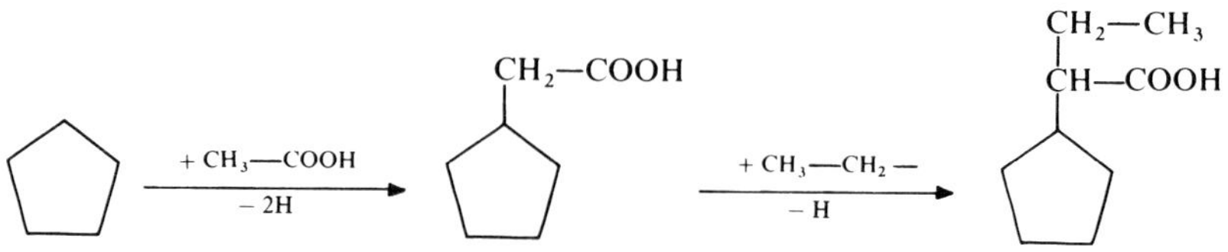

Cyclopentane   Cyclopentaneacetic acid   α-Ethylcyclopentaneacetic acid

R-1.2.4.2   **By placing a multiplicative prefix** [*such as 'bi-', 'ter-', etc. (see R-0.1.4.3)*], **before the name of the corresponding parent hydride (see R-2.4.4)**

Example:

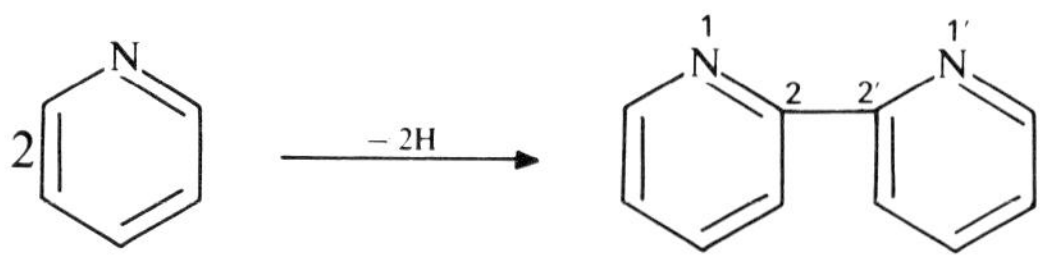

Pyridine   2,2′-Bipyridine

R-1.2.5   **Subtractive operation**

The subtractive operation involves the removal of an atom, ion, or group implicit in a name. This can occur with no other change, with exchange for hydrogen, with introduction of unsaturation, or with bond scission and reformation. The elimination of the elements of water with concomitant bond formation can also be regarded as a subtractive operation. Subtraction can be expressed in the following ways:

R-1.2.5.1   *By use of a prefix*

Examples:

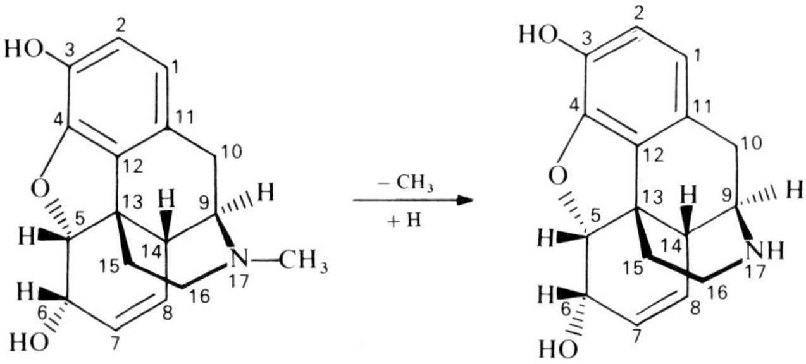

Morphine

Demethylmorphine
(exchange of methyl for H)

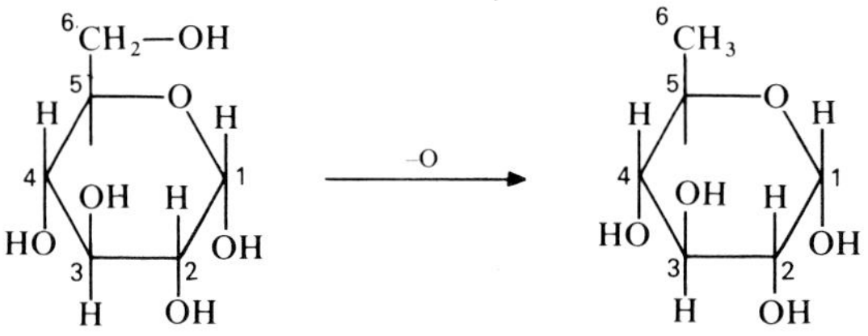

α-D-Glucopyranose

6-Deoxy-α-D-glucopyranose
(removal of O)

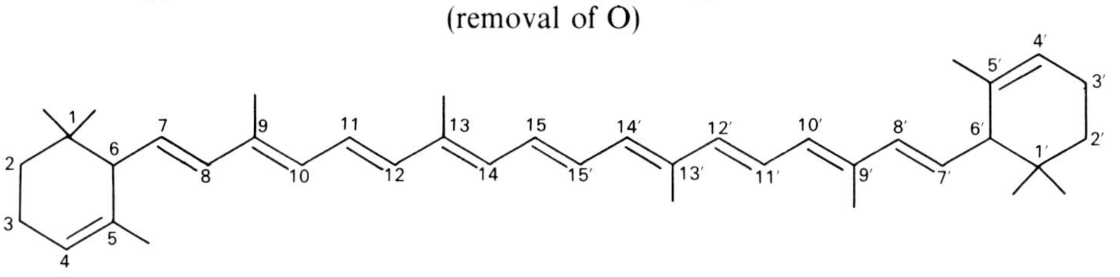

ε,ε-Carotene

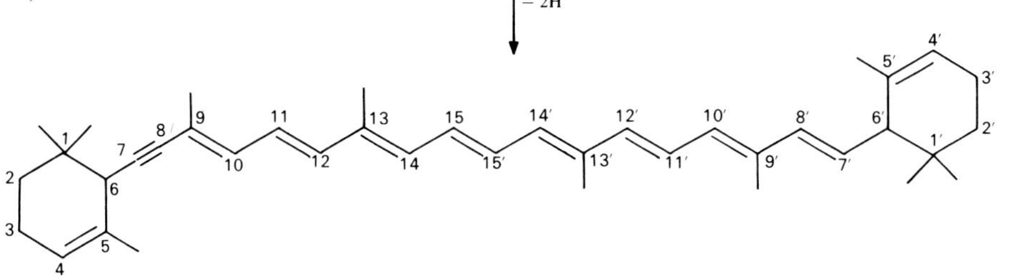

7,8-Didehydro-ε,ε-carotene

(formation of a C≡C bond from a −CH=CH− group by removal of two H atoms)

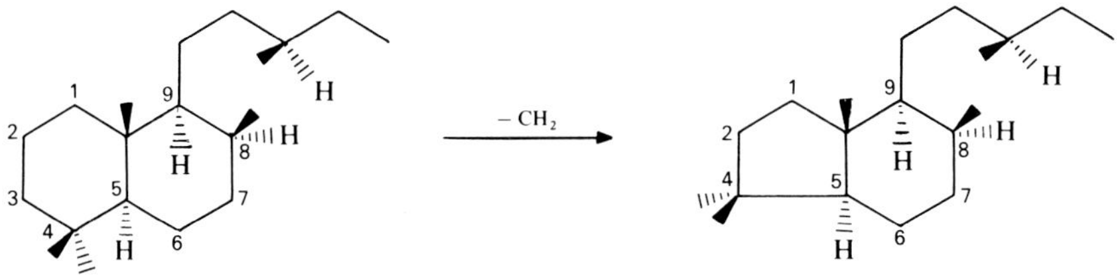

Labdane

3-Norlabdane
(ring contraction by removal of CH₂)

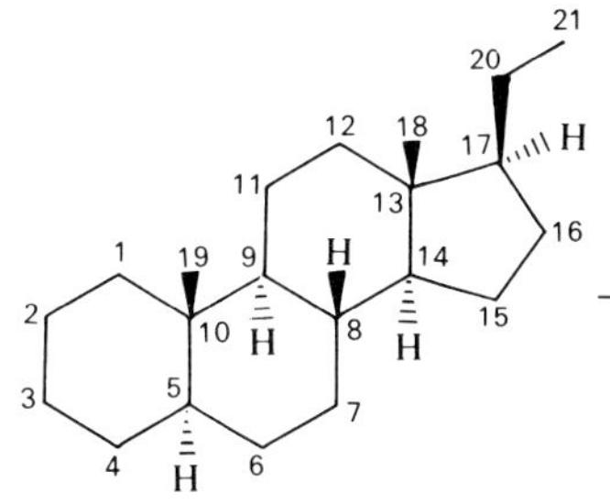

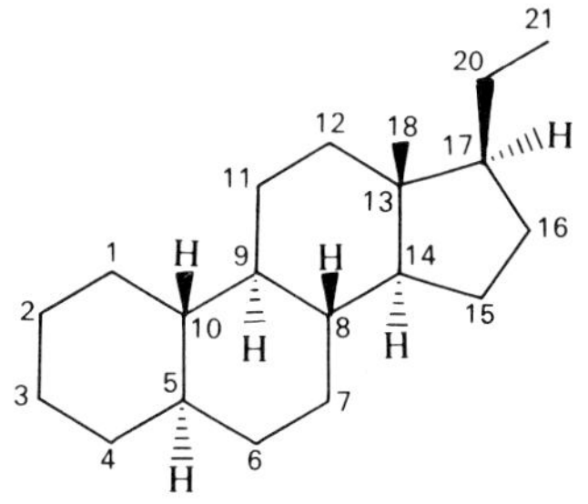

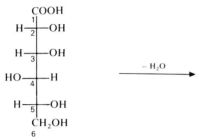

5α-Pregnane

19-Nor-5α-pregnane
(removal of CH₂ from a side chain)

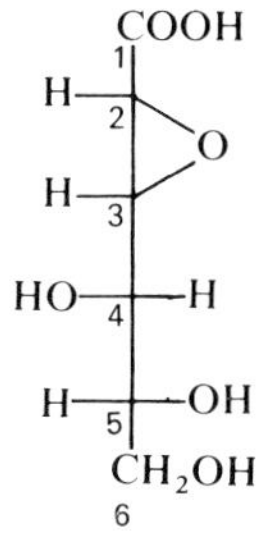

D-Gulonic acid

2,3-Anhydro-D-gulonic acid
(loss of H₂O from two hydroxy groups with bond formation)

Oxytocin ⟶ Des-7-proline-oxytocin
(here the prefix 'des-' signifies removal of the proline residue at position 7 of the polypeptide oxytocin, with rejoining of the chain) (see also Ref. 1, Section F).

R-1.2.5.2   **By a change in ending or addition of a suffix**

Examples:

$CH_3-[CH_2]_4-CH_3$ $\xrightarrow{-2H}$ $CH_3-CH_2-CH_2-CH=CH-CH_3$

Hexane

Hex-2-ene

Cycloheptane

Cyclohepta-1,3-diene

$CH_3-[CH_2]_3-CH_3$ $\xrightarrow{-6H}$ $CH\equiv C-CH_2-CH=CH_2$

Pentane

Pent-1-en-4-yne
('-en(e)' and '-yne' represent the loss of two and four H atoms, respectively, resulting in the introduction of unsaturation)

CH$_3$–CH$_3$ $\xrightarrow{\ -\text{H}^-\ }$ CH$_3$–CH$_2$$^+$

Ethane

Ethylium
('-ylium' indicates loss of H$^-$)

CH$_3$–CH$_2$–CH$_2$–CH$_3$ $\xrightarrow{\ -\text{H}^+\ }$ CH$_3$–CH$_2$–CH$_2$–CH$_2$$^-$

$\qquad\qquad\qquad\qquad\qquad\qquad\qquad\quad$ 4$\quad$ 3$\quad$ 2$\quad$ 1

Butane

Butan-1-ide
('-ide' indicates loss of H$^+$)

C$_6$H$_5$–SO$_3$H $\xrightarrow{\ -\text{H}^+\ }$ C$_6$H$_5$–SO$_3$$^-$

Benzenesulfonic acid

Benzenesulfonate
('-ate' indicates loss of H$^+$)

CH$_4$ $\xrightarrow{\ -\text{H}\cdot\ }$ CH$_3$·

Methane

Methyl
('-yl' indicates loss of H·)

R-1.2.6 Ring formation or cleavage

R-1.2.6.1 **The prefix cyclo-.** The formation of a ring by means of a direct link between any two atoms of a parent structure, with loss of one hydrogen atom from each, is indicated by the prefix 'cyclo-' followed by the name of the parent structure; when appropriate, this prefix is preceded by the locants of the positions joined by the new bond and by the Greek letter (α, β, γ) denoting the configurations at the ends of the new bond. (See R-2.3.1.1 and Section F of the 1979 edition of the IUPAC *Nomenclature of Organic Chemistry*[1]).

Examples:

CH$_3$–[CH$_2$]$_4$–CH$_3$ $\longrightarrow$

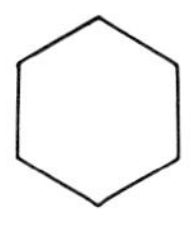

Hexane

Cyclohexane

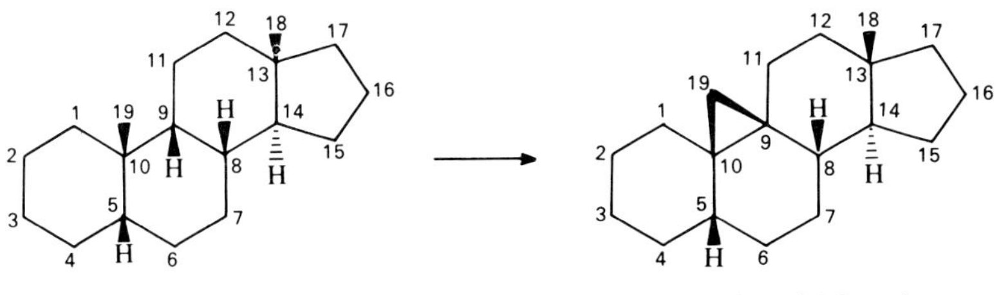

5β,9β-Androstane $\qquad\qquad$ 9,19-Cyclo-5β,9β-androstane

30

R-1.2.6.2   ***The prefix seco-.*** Cleavage of a ring with addition of one or more hydrogen atoms at each terminal group thus created is indicated by the prefix 'seco-' (see also Section F of the 1979 edition of the IUPAC *Nomenclature of Organic Chemistry*[1]).

Example:

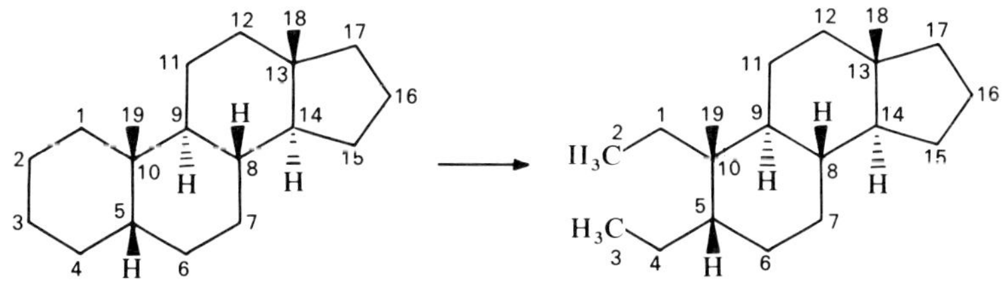

5β-Androstane   2,3-Seco-5β-androstane

R-1.2.7   **Rearrangement**

R-1.2.7.1   ***The prefix abeo-.*** Bond migration, i.e., the formal transfer of one end of a skeletal bond to another position, with a compensating transfer of a hydrogen atom, can be indicated by a prefix of the form '*x*(*y* → *z*)*abeo*-'. This prefix is compiled as follows: a numeral denoting the stationary (unchanged) end of the migrating bond (*x*) is followed by parentheses enclosing the locant denoting the original position (*y*) from which the other end of this bond has migrated, an arrow, and the locant (*z*) denoting the new position to which the bond has moved. The closing parenthesis is followed by the italicized prefix *abeo*- (Latin: I go away) to indicate bond migration. The original numbering is retained for the new compound and is used for the numbers *x*, *y*, and *z*. It is always necessary to specify the resulting stereochemistry (see also Section F of the 1979 edition of the IUPAC *Nomenclature of Organic Chemistry*[1]).

Example:

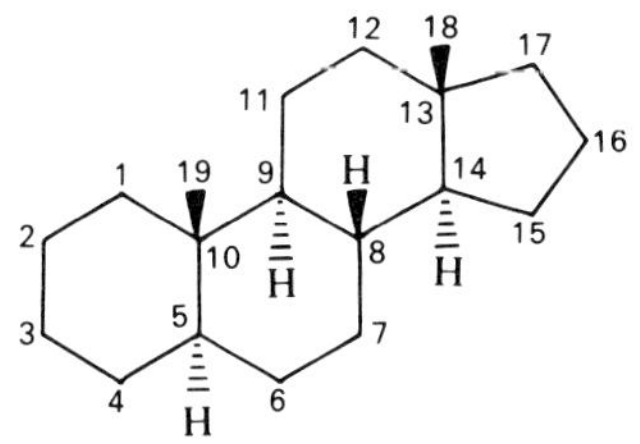

5α-Androstane

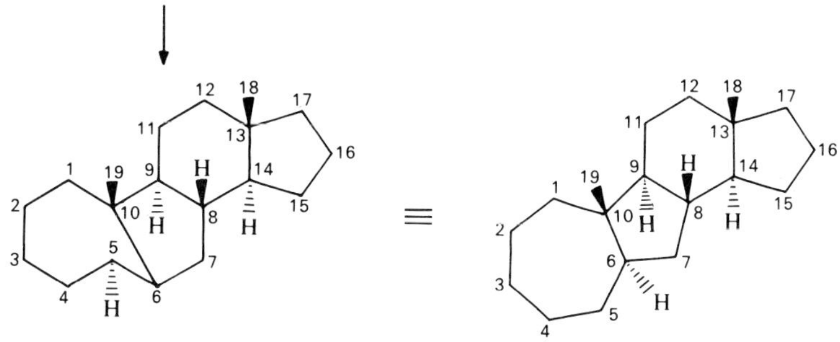

10(5 → 6) *abeo*-6α(H)-Androstane

R-1.2.7.2   ***The prefix retro-*** indicates a shift, by one position, of all single and double bonds in a conjugated polyene system. In this sense, it has been applied hitherto only to carotenoids (see Section F of the 1979 edition of the IUPAC *Nomenclature of Organic Chemistry*[1]).

Example:

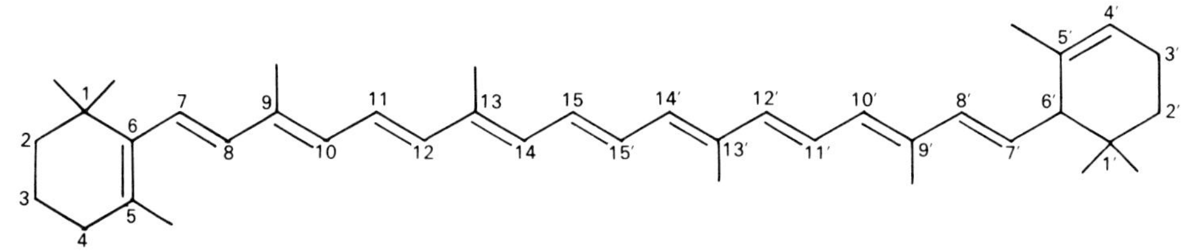

β,ε-Carotene

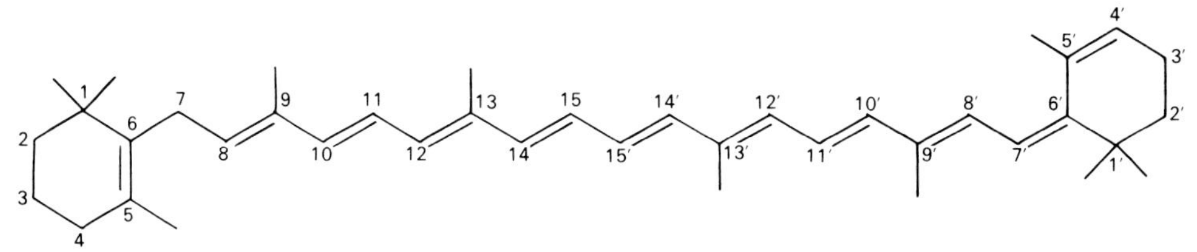

6',7-*retro*-β,ε-Carotene

Note: In naming peptides, this prefix is used with a different meaning, i.e., to indicate the change of Ala-Lys-Glu-Tyr-Leu to Leu-Tyr-Glu-Lys-Ala.

R-1.2.8   **Multiplicative operation**

R-1.2.8.1   ***Assemblies involving di- or polyvalent substituent groups.*** When a compound contains identical units whose only substituents are the principal characteristic groups, and when these identical units are linked by a symmetrical di- or polyvalent substituent group, it may be named by stating successively (a) the locants for the positions of substitution of the di- or polyvalent substituent group into the identical units, (b) the name of the di- or polyvalent substituent group, (c) the numerical prefix 'di-' or 'tri-', etc., and (d) the name of one of the identical units including the principal characteristic group. The numbering of the identical units and the principal characteristic group is retained, and when there is a further choice the points of substitution by the di- or polyvalent substituent group are numbered as low as possible. Primes, double primes, etc., are used to distinguish the locants of the identical units. If there is a choice, the greater number of primes is given to the locants of the unit having the higher numbered point of attachment to the di- or polyvalent group.

Examples:
S(CH$_2$–CH$_2$–COOH)$_2$
3,3'-Sulfanediyldipropanoic
  acid

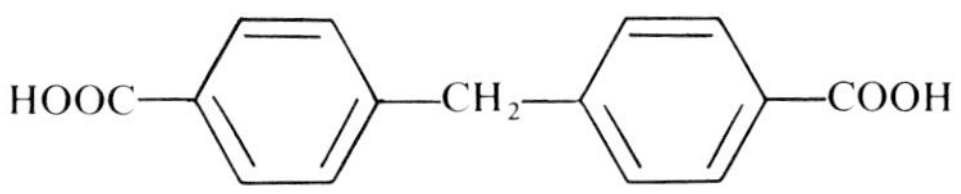

N(CH$_2$–COOH)$_3$
Nitrilotriacetic acid

HOOC—⟨benzene⟩—CH$_2$—⟨benzene⟩—COOH

4,4'-Methylenedibenzoic acid

32

R-1.2.8.2   ***Names of symmetrical di- or polyvalent substituent groups*** are formed by juxtaposing the names of the individual substituent groups starting with the central one, e.g., methylenedioxy, oxydiethylene; or the trivial name, if any, for compound substituent groups may be used.

Examples:

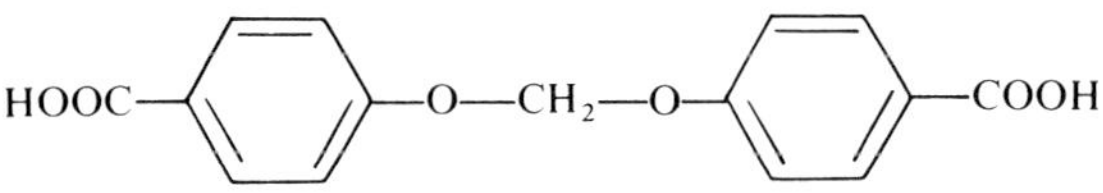

4,4'-(Methylenedioxy)dibenzoic acid

3,3',3'',3'''-[Oxybis(ethylenenitrilo)]tetrapropanoic acid

(In the last example, the compound substituent group name is formed by starting with the central substituent group 'oxy', followed by '-bis-', and adding successively the names of the substituent groups 'ethylene' ($-CH_2-CH_2-$) and 'nitrilo' ($-N<$), and finally the name of the unit 'propanoic acid', preceded by the multiplicative prefix 'tetra').

Note: In this nomenclature operation, unsymmetrical compound substituent groups are not used for linking identical units because of difficulty in assigning an unambiguous numbering to the complete structure. Instead, simple substitutive nomenclature is used.

Example:

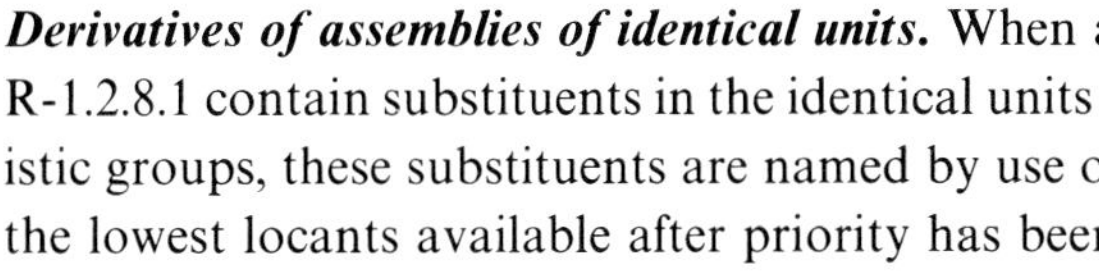

3-[2-(4-carboxyphenyl)ethoxy]benzoic acid

R-1.2.8.3   ***Derivatives of assemblies of identical units.*** When assemblies named in accordance with R-1.2.8.1 contain substituents in the identical units in addition to the principal characteristic groups, these substituents are named by use of prefixes. These prefixes are assigned the lowest locants available after priority has been given to the principal characteristic groups and the linking di- or polyvalent substituent groups.

Examples:

$N(CHCl-COOH)_3$
2,2',2''-Trichloronitrilotriacetic acid

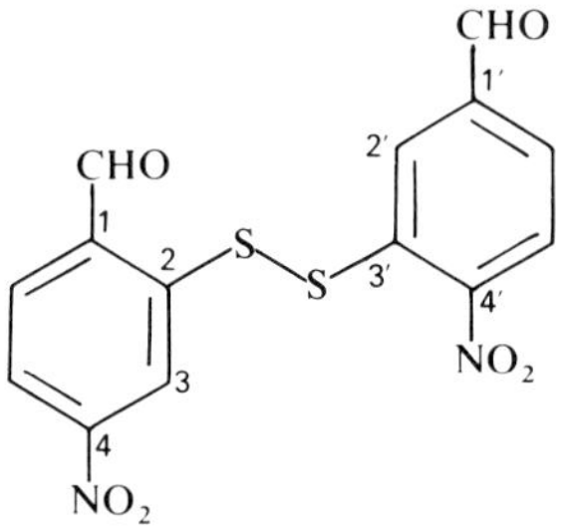

4,4'-Dinitro-2,3'-disulfanediyldibenzaldehyde 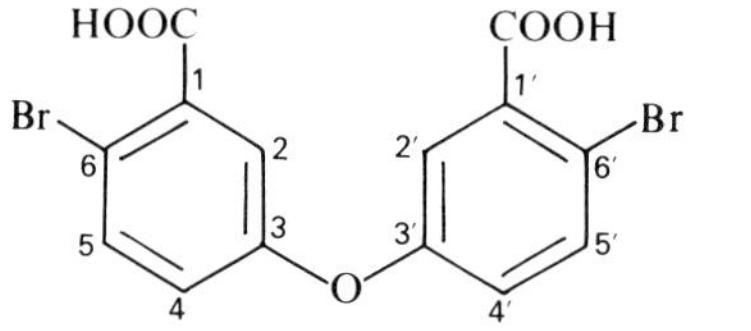   6,6'-Dibromo-3,3'-oxydibenzoic acid

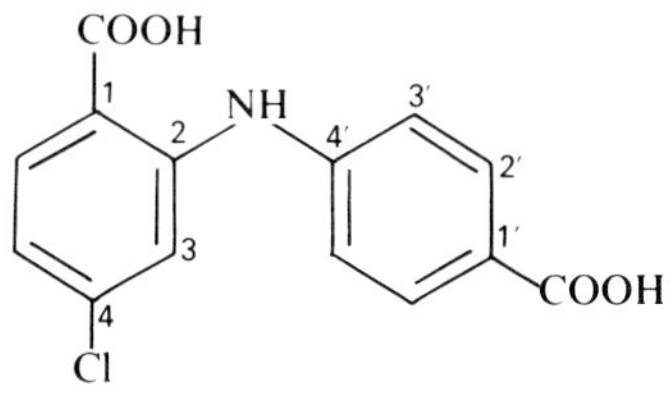

4-Chloro-2,4'-iminodibenzoic acid

Note: Assemblies of identical units linked by di- or polyvalent substituent groups where the individual units contain different numbers of the principal group are named by simple substitutive nomenclature.

R-1.3   INDICATED HYDROGEN

Under certain circumstances it is necessary to indicate in the name of a ring, or ring system, containing the maximum number of noncumulative double bonds, one or more positions where no multiple bond is attached. This is done by specifying the presence of an 'extra' hydrogen atom at such positions by citation of the appropriate numerical locant followed by an italicized capital *H*.

Example:

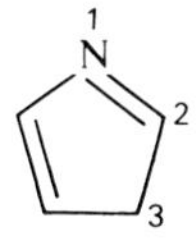

3*H*-Pyrrole

In the above example, the 'indicated hydrogen' locates two hydrogen atoms at position 3 (an 'extra' hydrogen over the number present if there were a double bond in the ring at that position), thus specifying a particular pyrrole tautomer, as illustrated. Indicated hydrogen of this type normally precedes the name of a parent hydride.

A second type of indicated hydrogen (sometimes referred to as 'added hydrogen') describes hydrogen atoms added to a specified structure as a consequence of the addition of a suffix or a prefix describing a structural modification. This type of indicated hydrogen is normally cited in parentheses after the locant of the additional structural feature.

34

Example:

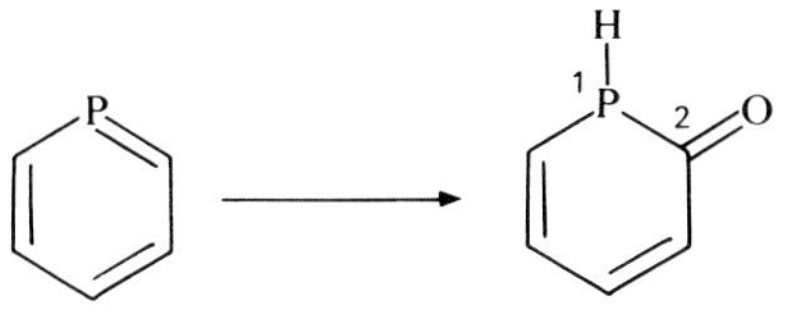

Phosphinine   Phosphinin-2(1*H*)-one[17]

The detailed procedures for using indicated hydrogen, including differences in approach used in index nomenclature by Chemical Abstracts Service and in *Beilstein*, will be described in a separate publication.

---

[17] This compound may also be named as 1,2-dihydrophosphinin-2-one, in which the hydro prefixes are nondetachable. Rule C-316.1 of the IUPAC *Nomenclature of Organic Chemistry*[1], leading to the name 2-oxo-1,2-dihydrophosphinine, is now abandoned.

R-2   Parent Hydrides[18] and their Derived Substituent Groups

R-2.0   INTRODUCTION

A 'parent hydride' (see R-0.2.1.1) is the structure which is named before the addition of affixes denoting substituents to yield the name of a specific compound. Its name is understood to signify a definite population of hydrogen atoms attached to a skeletal structure. Acyclic parent hydrides are always unbranched, for example pentane and trisilane; although a few trivial names for branched acyclic hydrocarbons are retained (see R-9.1, Table 19(a), p. 163), their use for naming substitutive derivatives is not recommended. Cyclic parent hydrides are usually either fully saturated, for example, cyclopentane, cyclotrisiloxane, azepane, bicyclo[2.2.2]octane, and spiro[4.5]decane, or fully unsaturated, i.e., they contain the maximum number of noncumulative double bonds, for example, pyridine, 1,3-oxazole, 1H-phenalene, phenanthroline, and benzo[a]-anthracene. Also, there are parent hydrides that are partially saturated, for example, 1,4-dihydro-1,4-ethanoanthracene and spiro[1,3-dioxolane-2,1'-[1H]indene], and that are combinations of cyclic and saturated acyclic structures having retained trivial names (see R-9.1, Table 19(a), p. 163).

R-2.1   MONONUCLEAR HYDRIDES

The names of mononuclear hydrides of the elements for use as parents in substitutive nomenclature are given in Table 2. Many are formed systematically by combining the 'a' term of the element (with elision of the terminal 'a') with the ending '-ane', for example, borane for BH_3, silane for SiH_4, etc. There are important exceptions: methane for CH_4, oxidane for OH_2, sulfane for SH_2, selane for SeH_2, etc., (see Table 2). The systematic alternatives to well established common names, e.g., azane for ammonia, oxidane for water, and chlorane, bromane, etc., for hydrogen chloride, bromide, etc., are necessary for naming some derivatives and for generating names of polynuclear homologues. If the bonding number of the element differs from the normal one as defined in R-1.1, the name of the hydride is modified by affixing a λ^n symbol as instructed by R-1.1.

R-2.2   ACYCLIC POLYNUCLEAR HYDRIDES

R-2.2.1   **Hydrocarbons**

The saturated unbranched acyclic hydrocarbons from C_2 to C_4 are named ethane, propane, and butane (see R-9.1, Table 19(a), p. 162). Systematic names of the higher members of this series consist of a numerical term (see Table 11, p. 71), followed by '-ane' with elision of a terminal 'a' from the basic numerical term. The generic name for saturated acyclic hydrocarbons (branched or unbranched) is 'alkane'. The chain is numbered from one end to the other with arabic numerals.

---

[18] More detailed recommendations for naming these compounds are given in Sections A, B, and D of the 1979 edition of the IUPAC *Nomenclature of Organic Chemistry*[1].

Table 2 Mononuclear hydrides

BH_3	Borane[a]	OH_2	Oxidane[c,e,f]
CH_4	Methane[b] (Carbane)	SH_2	Sulfane[e,f]
SiH_4	Silane	SH_4	λ^4-Sulfane[e,g]
GeH_4	Germane	SH_6	λ^6-Sulfane[e,g]
SnH_4	Stannane	SeH_2	Selane[e,f,g]
PbH_4	Plumbane	TeH_2	Tellane[e,f]
NH_3	Azane[c,d]	PoH_2	Polane[e,f]
PH_3	Phosphane[b] (Phosphine)	FH	Fluorane[c,f]
PH_5	λ^5-Phosphane[b] (Phosphorane)	ClH	Chlorane[c,f]
AsH_3	Arsane[b] (Arsine)	BrH	Bromane[c,f]
AsH_5	λ^5-Arsane[b] (Arsorane)	IH	Iodane[c,f]
SbH_3	Stibane[b] (Stibine)	IH_3	λ^3-Iodane[h]
SbH_5	λ^5-Stibane[b] (Stiborane)	IH_5	λ^5-Iodane[g]
BiH_3	Bismuthane[b,e] (Bismuthine)	AtH	Astatane[f]

[a] Although analogous names have been suggested for aluminium and gallium hydrides, they are not included in these recommendations.

[b] Preferred name.

[c] 'Ammonia', 'water', 'hydrogen chloride' (and analogues) are common names for NH_3, OH_2, and ClH (and analogues), respectively.

[d] The name amine is used, for example, in *Beilstein*, as a parent hydride name for NH_3.

[e] Exceptions. The expected names, bismane, oxane, thiane, selenane, tellurane and polonane are Hantzsch–Widman names for heteromonocyclic rings (see R-2.3.3).

[f] The normal formulae H_2O, H_2S, etc., have been reversed in this table for the purpose of comparison.

[g] The names 'sulfurane' and 'selenurane', and related names, such as persulfurane and periodinane, which have been used in recent literature, are not recommended.

[h] The name 'iodinane', which has been used in recent literature, cannot be used because it is a Hantzsch–Widman name for a heteromonocyclic iodine ring (see R-2.3.3).

Examples:

$$CH_3-CH_2-CH_2-CH_2-CH_2-CH_2-CH_3$$
$$7\quad 6\quad 5\quad 4\quad 3\quad 2\quad 1$$

Heptane

$$CH_3-[CH_2]_{58}-CH_3$$
$$60\qquad\qquad\qquad 1$$

Hexacontane

$$CH_3-[CH_2]_{21}-CH_3$$
$$23\qquad\qquad 1$$

Tricosane

R-2.2.2   **Homogeneous hydrides other than hydrocarbons or boron hydrides[19]**

A compound consisting of an unbranched chain containing several identical hetero-atoms saturated with hydrogen atoms, may be named by citing the appropriate multiplying prefix (with *no elision* of the terminal vowel of the multiplying prefix) followed by the appropriate name of the hydride according to R-2.1. If necessary, a λ^n symbol is used, according to R-1.1.

---

[19] Polyboranes are not covered in these recommendations; because of their unique bonding characteristics, they have been treated in IUPAC *Nomenclature of Inorganic Chemistry*[3], Recommendations I-11.1 through I-11.3, pp. 207–225.

Examples:

NH$_2$–NH$_2$
  2   1

Diazane (or Hydrazine)

SiH$_3$–SiH$_2$–SiH$_2$–SiH$_2$–SiH$_3$
  5   4   3   2   1

Pentasilane

PH$_2$–PH–PH–PH–PH$_2$
  5   4   3   2   1

Pentaphosphane

SH–SH$_4$–SH
  3   2   1

2λ^6-Trisulfane

H$_2$N–[NH]$_7$–NH$_2$
         9   1

Nonaazane

R-2.2.3 Heterogeneous hydrides

R-2.2.3.1 Heterogeneous hydrides consisting of chains containing carbon atoms and several heteroatoms, alike or different, and terminating with carbon may be named by replacement nomenclature (see R-1.2.2).

Example:

CH$_3$–O–CH$_2$–O–CH$_2$–CH$_2$–CH$_2$–O–CH$_2$–O–CH$_3$
 11   10 9   8   7   6   5   4   3   2   1

2,4,8,10-Tetraoxaundecane

R-2.2.3.2 Compounds containing an unbranched chain of alternating atoms terminated by two identical atoms of the element coming later in Table 3 (see R-2.3.3.3) may be named by citing successively a multiplying prefix denoting the number of atoms of the terminating element followed by the 'a' term of that element (see R-9.3), then the 'a' term of the other element of the chain and the ending '-ane'. The terminal letter 'a' of an 'a' term is elided when followed by a vowel; the terminal vowel of a numerical prefix is not elided even when the 'a' term begins with the same vowel.

Examples:

SiH$_3$–NH–SiH$_3$
  3   2   1

Disilazane

SnH$_3$–O–SnH$_2$–O–SnH$_2$–O–SnH$_3$
  7   6   5   4   3   2   1

Tetrastannoxane

AsH$_2$–NH–AsH–NH–AsH–NH–AsH$_2$
  7   6   5   4   3   2   1

Tetraarsazane

38

R-2.3   MONOCYCLIC HYDRIDES

R-2.3.1   **Hydrocarbons**

R-2.3.1.1   The names of saturated monocyclic hydrocarbons are formed by attaching the prefix
             'cyclo-' to the name of the acyclic saturated unbranched hydrocarbon with the same
             number of carbon atoms. The generic name of monocyclic hydrocarbons is 'cycloalkane'.
             Numbering proceeds sequentially round the ring.

             Examples:

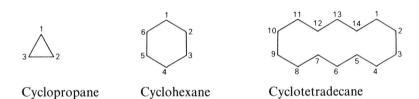

             Cyclopropane   Cyclohexane   Cyclotetradecane

R-2.3.1.2   Unsubstituted monocyclic hydrocarbon polyenes having the maximum number of
             noncumulative double bonds and with a general formula of C_nH_n or C_nH_{2n+1} (with n
             greater than 6) are called annulenes generically[20]. A specific annulene may be named as
             an [n]annulene, where n is the number of carbon atoms of the ring. When n is an odd
             number, i.e., when the annulene has the general formula C_nH_{n+1} the extra hydrogen
             atom is denoted as 'indicated hydrogen' (see R-1.3).

             Examples:

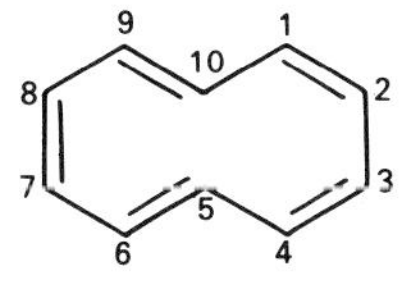

             [10]Annulene   1H-[9]Annulene
             (Cyclodecapentaene   (Cyclonona-1,3,5,7-tetraene
             according to R-3.1.1)   according to R-3.1.1)

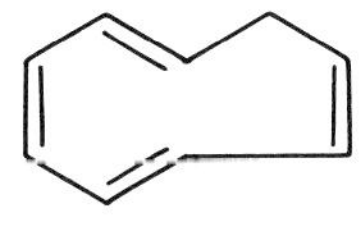

R-2.3.2   **Homogeneous hydrides other than hydrocarbons or boron hydrides** (see footnote 19)
             A compound having a single saturated ring of identical heteroatoms is named by adding
             the prefix 'cyclo-' to the name of the saturated unbranched chain containing the same
             number of identical atoms[21].

---

[20] When $n = 6$, the name is benzene (see R-9.1, Table 19(a), p. 162)
[21] Such rings with 10 or fewer members may also be named according to the Hantzsch–Widman system (see
R-2.3.3).

Examples:

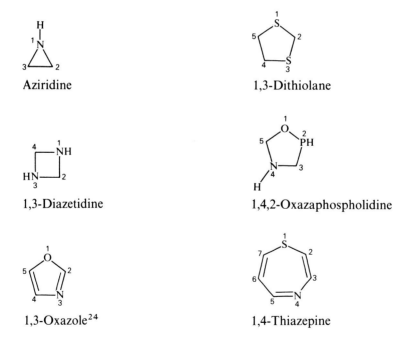

R-2.3.3   **Heterogeneous hydrides other than heteropolyboron hydrides[22]**

R-2.3.3.1   Monocyclic compounds with no more than 10 ring members and containing one or more heteroatoms, may be named by using the extended Hantzsch–Widman system[23]. The name is formed by combining the 'a' prefix(es) for the heteroatom(s) (Table 3) with a stem indicating the size of the ring (Table 4). The prefixes designating heteroatoms are cited in the order of their appearance in Table 3; and the heteroatoms of the structure are numbered in the same order.

Examples of Hantzsch–Widman names:

----

[22] Heteropolyboranes are not covered in these recommendations; because of their unique bonding characteristics, they have been treated in *IUPAC Nomenclature of Inorganic Chemistry*[3], Recommendations I-11.4.3.1 through I-11.4.3.3, pp. 228–232.
[23] International Union of Pure and Applied Chemistry. Organic Chemistry Division. Commission on Nomenclature of Organic Chemistry, 'Revision of the Extended Hantzsch–Widman System of Nomenclature for Heteromonocycles', *Pure Appl. Chem.*, **55**, 409–416 (1983).
[24] Even though the trivial name isoxazole and its chalcogen analogues are retained (see R-9.1, Table 23, p. 168), it is recommended that the locants 1,3- be always used with oxazole and its chalcogen analogues to form a proper Hantzsch–Widman name.

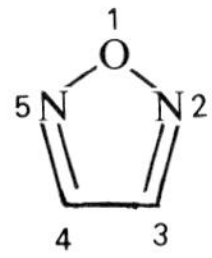

1,2,5-Oxadiazole
(formerly Furazan)

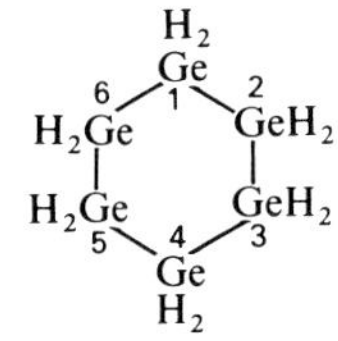

Hexagerminane
(Cyclohexagermane, according to R-2.3.2)

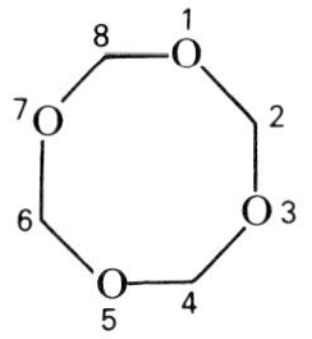

1,3,5,7-Tetroxocane

Retained trivial and semisystematic names for heteromonocyclic compounds are given in R-9.1, Tables 23, p. 166 and 24, p. 171.

R-2.3.3.1.1   The position of a single heteroatom determines the numbering in a monocyclic compound.

Example:

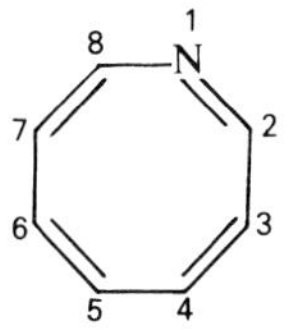

Azocine

R-2.3.3.1.2   When the same heteroatom occurs more than once in a ring, the numbering is chosen to give the lowest locants to the heteroatoms.

Example:

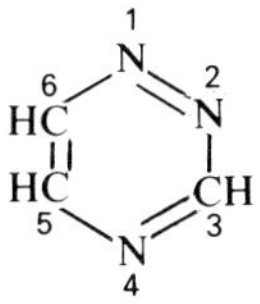

1,2,4-Triazine

R-2.3.3.1.3   When heteroatoms of different kinds are present, the locant 1 is given to a heteroatom which appears earliest in Table 3. The numbering is then chosen to give the lowest locants to the heteroatoms, first considered as a set without regard to kind; if a choice still remains, then to a heteroatom appearing earliest in Table 3.

Table 3 Hantzsch–Widman system prefixes (in decreasing order of priority)

Element	Bonding number (valence)		Prefix	Element	Bonding number (valence)		Prefix
Fluorine	1	(I)	Fluora	Arsenic	3	(III)	Arsa
Chlorine	1	(I)	Chlora	Antimony	3	(III)	Stiba
Bromine	1	(I)	Broma	Bismuth	3	(III)	Bisma
Iodine	1	(I)	Ioda	Silicon	4	(IV)	Sila
Oxygen	2	(II)	Oxa	Germanium	4	(IV)	Germa
Sulfur	2	(II)	Thia	Tin	4	(IV)	Stanna
Selenium	2	(II)	Selena	Lead	4	(IV)	Plumba
Tellurium	2	(II)	Tellura	Boron	3	(III)	Bora
Nitrogen	3	(III)	Aza	Mercury	2	(II)	Mercura
Phosphorus	3	(III)	Phospha				

Table 4 Hantzsch–Widman system stems

Ring size	Unsaturated[b]	Saturated[c]	Ring size	Unsaturated[b]	Saturated[c]
3	irene[d]	irane[e]	7	epine	epane
4	ete	etane[e]	8	ocine	ocane
5	ole	olane[e]	9	onine	onane
6A[a]	ine	ane	10	ecine	ecane
6B[a]	ine	inane			
6C[a]	inine	inane			

[a] The stem for *six-membered rings* depends on the least preferred heteroatom in the ring, i.e., the heteroatom whose name directly precedes the stem. To determine the proper stem for a six-membered ring, select the set below that contains the least preferred heteroatom before consulting the table. For example, the proper stem for the dioxazine ring is found after set 6B, which contains the element nitrogen.

6A: O, S, Se, Te, Bi, Hg
6B: N, Si, Ge, Sn, Pb
6C: B, F, Cl, Br, I, P, As, Sb

[b] Used when the ring contains the maximum number of noncumulative double bonds and at least one double bond is present when the heteroatoms have the bonding numbers (valences) given in Table 1 (see Note 2, below).
[c] Used when no double bonds are present or when none is possible (see Note 2, below).
[d] The traditional stem 'irine' may be used for rings containing only nitrogen.
[e] The traditional stems 'iridine', 'etidine', and 'olidine' are preferred for rings containing nitrogen and are used for saturated heteromonocycles having three, four, or five ring members, respectively.

Note 1 Prefixes for the halogen elements are included in order to provide for naming heteromonocycles containing cationic halogen atoms (*Pure Appl. Chem.* **65**, 1357–1455 (1993)) and halogen atoms in nonstandard valency states.
Note 2 Although a terminal 'e' is used on all stems in these recommendations, it is optional. (Stems without the terminal 'e' for unsaturated nonnitrogenous rings with six or more ring members are used in CAS index nomenclature, for example, dioxin, and dioxaphosphepin).
Note 3 The stems for ring sizes 3, 4, 7, 8, 9, and 10 may be considered to be derived from numerical prefixes as follows: 'ir' from *tri*, 'et' from *tetra*, 'ep' from *hepta*, 'oc' from *octa*, 'on' from *nona*, and 'ec' from *deca*.
Note 4 The stems 'etine' and 'oline', which would be consistent with the other stems for unsaturated rings, cannot be used because they were formerly used to name nitrogenous four- and five-membered rings, respectively, having only one of two possible double bonds.
Note 5 Oxine must not be used for pyran because it has been used as a trivial name for quinolin-8-ol.
Note 6 Azine must not be used for pyridine because of its long-established use as a class name for =N–N= compounds (see R-5.6.6.3).

Examples:

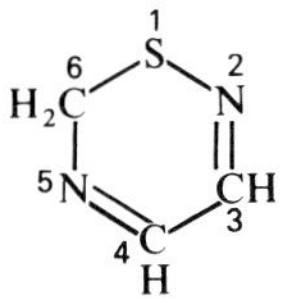

6*H*-1,2,5-Thiadiazine
(**not**: 2,1,4-Thiadiazine)
(**not**: 1,3,6-Thiadiazine)

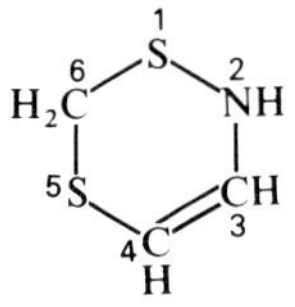

2*H*,6*H*-1,5,2-Dithiazine
(**not**: 1,3,4-Dithiazine)
(**not**: 1,3,6-Dithiazine)
(**not**: 1,5,4-Dithiazine)

The numbering must begin with the sulfur atom. This condition eliminates 2,1,4-thiadiazine and the set of locants 1,2,5 is preferred to 1,3,6.

The numbering has to begin with a sulfur atom. The choice of which sulfur atom will begin the numbering is determined by the lower set of locants of the remaining heteroatoms without regard to kind.
As the set 1,2,5 is lower than 1,3,4 or 1,3,6 or 1,5,4 in the usual sense, the name 1,5,2-dithiazine is preferred.

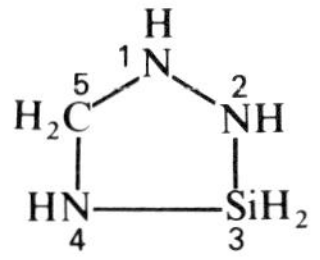

1,2,4,3-Triazasilolidine
(for the N atoms, the set 1,2,4 is preferred to 1,3,4).

R-2.3.3.2   Heteromonocyclic compounds may be named by replacement nomenclature. However, replacement nomenclature for heteromonocycles with 10 or fewer members has usually been applied only to silicon-containing rings. Numbering follows R-2.3.3.1.

Examples:

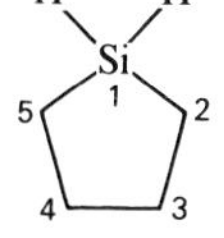

Silacyclopentane 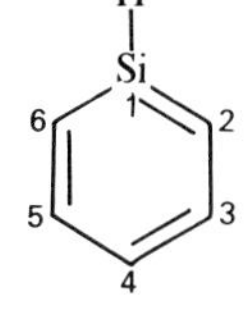   Silabenzene 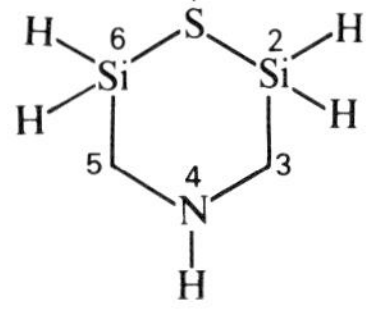   1-Thia-4-aza-2,6-disilacyclohexane

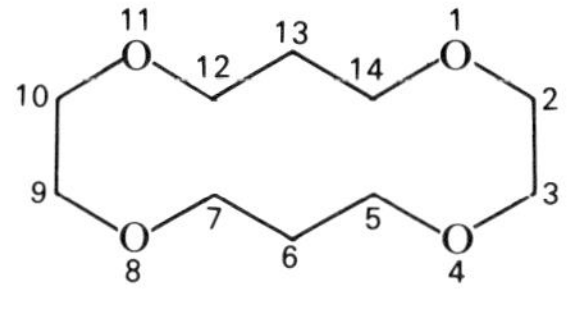

1,4,8,11-Tetraoxacyclotetradecane

R-2.3.3.3   Saturated monocyclic systems consisting of repeating units of two different skeletal atoms may be named by citing successively the prefix 'cyclo-' followed by a multiplying infix denoting the number of repeating units, the 'a' terms of the atoms of the repeating unit in the reverse order to that given in R-9.3, and the suffix '-ane'. The terminal letter 'a' of an 'a' term is elided when followed by a vowel; the terminal vowel of a numerical prefix is not elided even when the 'a' term begins with the same vowel. Numbering follows, R-2.3.3.1.

Examples:

Cyclotetraazoxane   Cyclotriboraphosphane

R-2.4   **POLYCYCLIC PARENT HYDRIDES**
             Polycyclic ring systems may be designated by trivial or semisystematic names (see R-9.1, Tables 20, p. 164, 21, p. 166, 23, p. 166, and 24, p. 171). Replacement nomenclature may be used in a similar manner to that for heteromonocyclic compounds (see R-2.3.3). Several systems are used to name polycyclic ring systems.

R-2.4.1   **Fusion nomenclature**
             Polycyclic ring systems in which any two adjacent rings have two, and only two, adjacent atoms in common are said to be '*ortho*-fused'. Such ring systems have *n* common sides and *2n* common atoms. Polycyclic ring systems in which a ring contains two, and only two, adjacent atoms in common with each of two or more rings of a contiguous series of *ortho*-fused rings are said to be '*ortho*- and *peri*-fused'. Such ring systems have *n* common sides and fewer than *2n* common atoms.

Examples:

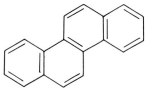

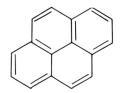

3 common sides   5 common sides
6 common atoms   6 common atoms
an '*ortho*-fused' system   an '*ortho*- and *peri*-fused' system

R-2.4.1.1   *ortho*-Fused or *ortho*- and *peri*-fused polycyclic ring systems with the maximum number of noncumulative double bonds and which have no accepted trivial or semisystematic name, are named by selecting a component having a trivial or semisystematic name as a *principal* component and designating the other components as prefixes. The principal component should be a heterocyclic system (if possible) and should include as many rings as possible.

The prefixes designating an attached component are formed by changing the terminal 'e' of a trivial or Hantzsch–Widman name (R-2.3.3) of a component into 'o'. In these recommendations, this final 'o' is no longer elided when followed by a vowel. Contracted fusion prefixes such as 'benzo-', 'naphtho-', and 'anthra-' are used. For monocyclic components other than benzo-, prefix names such as 'cyclopenta-' and 'cyclohepta-' are used and represent the form having the maximum number of noncumulative double bonds. Locants referring to positions of heteroatoms in component rings are enclosed in brackets. Further details of the principles of fusion nomenclature are given in Section A, pp. 22–29 and in Section B, pp. 64–68 in the 1979 edition of the IUPAC *Nomenclature of Organic Chemistry*[1].

Examples:

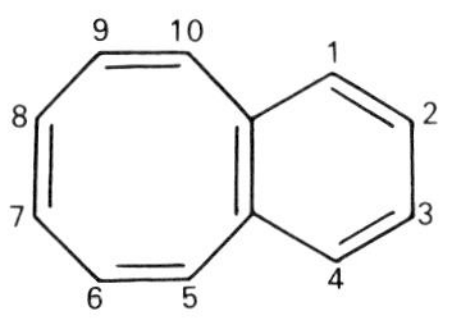

Benzo[8]annulene (see R-2.3.1.2)[25]

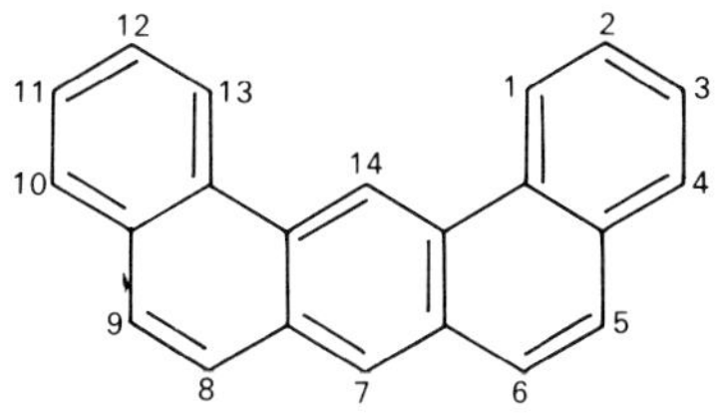

Dibenzo[*a,j*]anthracene

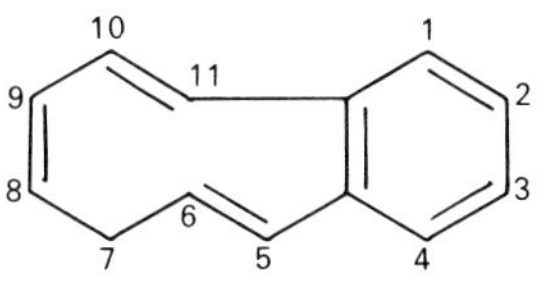

7*H*-Benzo[9]annulene[26]

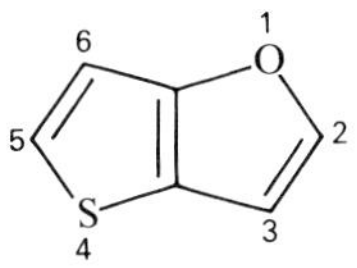

Thieno[3,2-*b*]furan

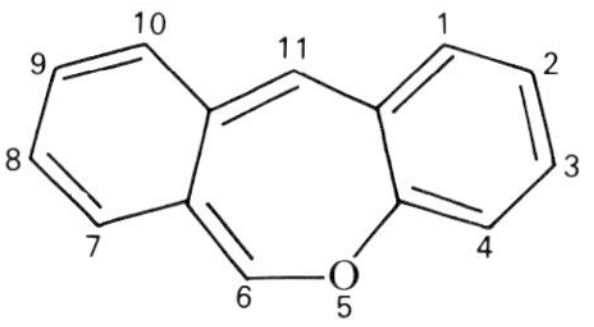

Dibenzo[*b,e*]oxepine

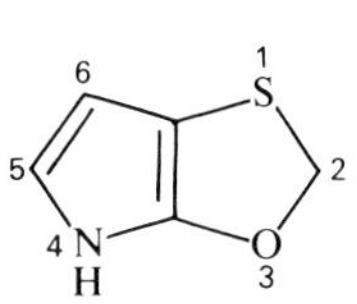

4*H*-[1,3]Oxathiolo[5,4-*b*]pyrrole

---

[25] The name 'benzocyclooctene' attributed to this compound on the basis of the previous edition of the IUPAC *Nomenclature of Organic Chemistry*[1] could be ambiguous because, according to Rule A-21.4, the name benzocyclooctene describes a two-component ring system containing the maximum number of noncumulative double bonds while, according to Rule A-23.5, it would describe its 5,6,7,8,9,10-hexahydro derivative. Its use is no longer encouraged and the Commission recommends names based on [*n*]annulene.

[26] As in the previous footnote, the name '7*H*-Benzocycloheptene' derived on the basis of Rule A-21.5 of the IUPAC *Nomenclature of Organic Chemistry*[1] could be ambiguous and is no longer encouraged.

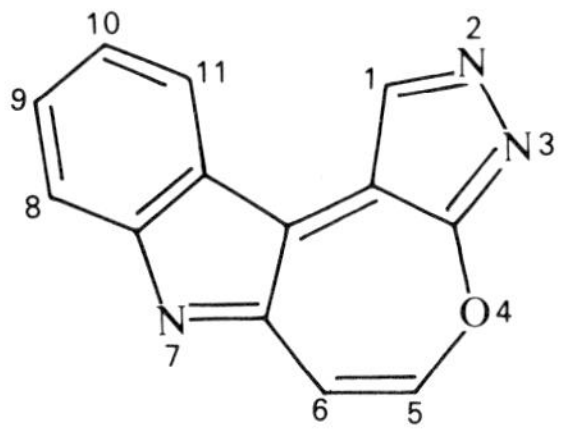

Pyrazolo[4',3':6,7]oxepino[4,5-*b*]indole

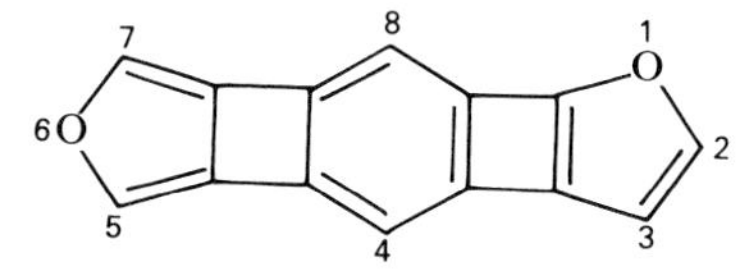

Benzo[1'',2'':3,4;4'',5'':3',4']dicyclobuta-[1,2-*b*:1'2'-*c*']difuran

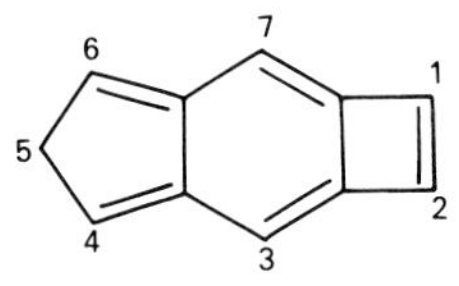

5*H*-Cyclobuta[*f*]indene

R-2.4.1.2   A polycyclic system which can be regarded as '*ortho*-fused' or '*ortho*- and *peri*-fused' and which, at the same time, has other bridges can be named by citing the bridges as nondetachable prefixes to the name of the '*ortho*-fused' or '*ortho*- and *peri*-fused' system.

Examples:

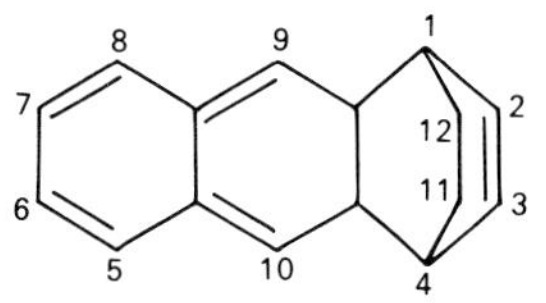

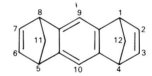

1,4-Dihydro-1,4-ethanoanthracene   1,4,5,8-Tetrahydro-1,4:5,8-dimethanoanthracene

Note: A double bond in a bridge is denoted by a locant enclosed in square brackets (see R-9.2.1).

Example:

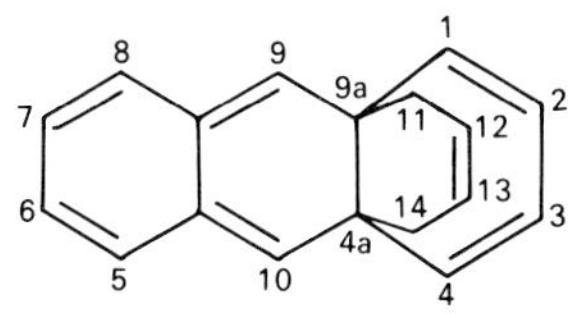

4a,9a-But[2]enoanthracene

R-2.4.1.3   There are several methods that can be used to derive names for polycyclic components of fused hydrocarbon ring systems.

R-2.4.1.3.1 Hydrocarbons consisting of four or more *ortho*-fused benzene rings in a rectilinear arrangement are named by citing a numerical prefix denoting the number of benzene rings followed by the term '-acene' with elision of an 'a'[27].

---

[27] The name tetracene is now preferred to naphthacene, the name used previously for the polycyclic hydrocarbon with four *ortho*-fused benzene rings in a rectilinear arrangement.

Example:

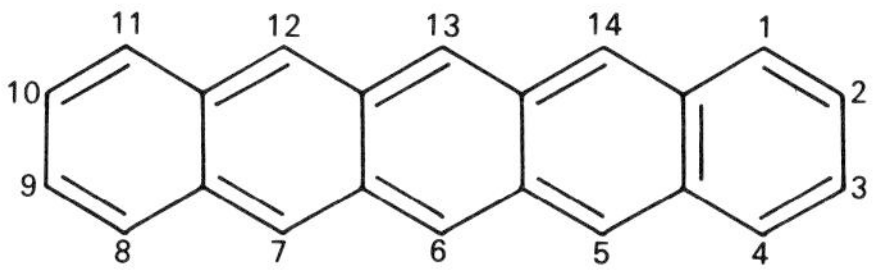

Pentacene

R-2.4.1.3.2   Hydrocarbons consisting of a benzene ring *ortho*-fused at positions 1,2 and 3,4 to two identical rectilinear arrangements of *ortho*-fused benzene rings (or to two rectilinear arrangements of *ortho*-fused benzene rings one of which has one more ring than the other) are named by citing the numerical prefix denoting the number of benzene rings followed by the term '-aphene' with elision of an 'a'.

Examples:

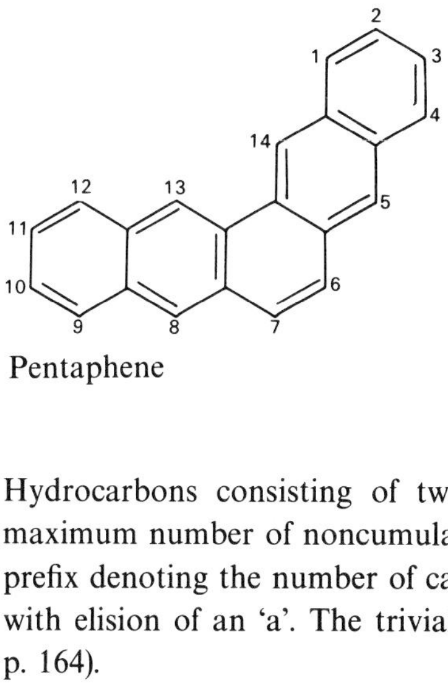

Pentaphene 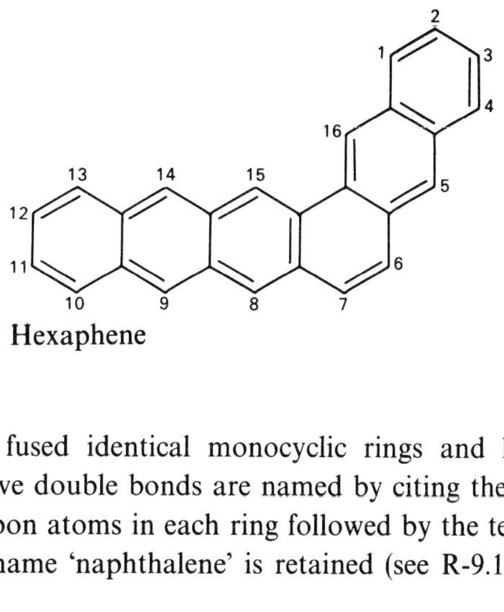   Hexaphene

R-2.4.1.3.3   Hydrocarbons consisting of two fused identical monocyclic rings and having the maximum number of noncumulative double bonds are named by citing the numerical prefix denoting the number of carbon atoms in each ring followed by the term '-alene' with elision of an 'a'. The trivial name 'naphthalene' is retained (see R-9.1, Table 20, p. 164).

Examples:

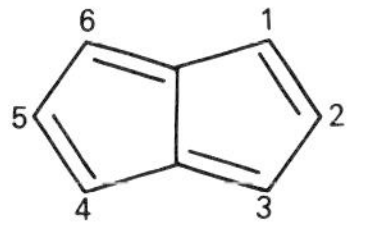

Pentalene 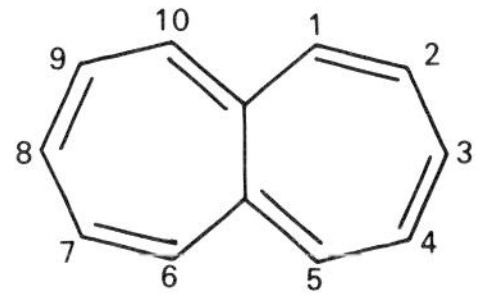   Heptalene 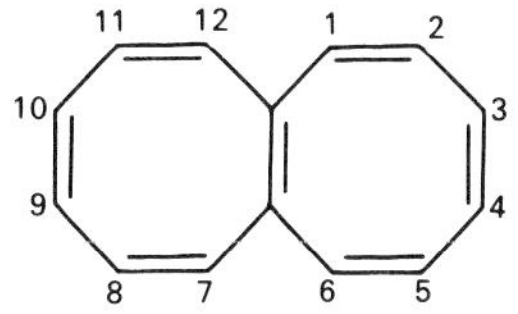   Octalene

R-2.4.1.3.4   Hydrocarbons consisting of a monocyclic hydrocarbon with an even number of carbon atoms *ortho*-fused on alternate sides to benzene rings are named by citing a numerical prefix denoting the number of benzene rings followed by the term '-phenylene'.

Examples:

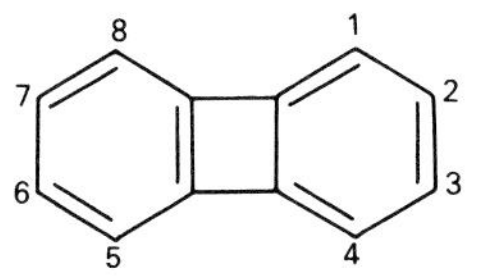

Biphenylene (**not** Diphenylene) 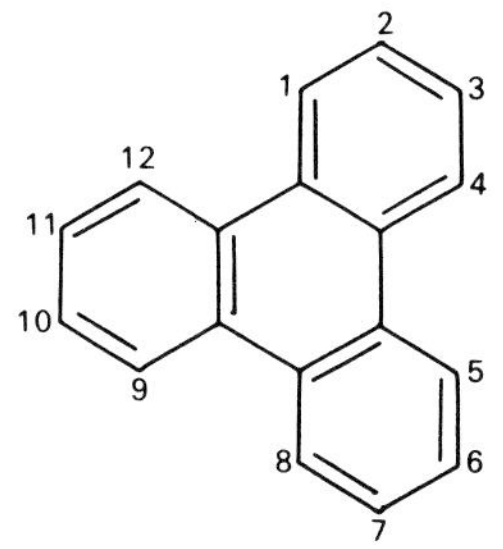   Triphenylene

R-2.4.1.3.5   Hydrocarbons consisting of a monocyclic hydrocarbon with an even number of carbon atoms *ortho*-fused on alternate sides to the 2,3-positions of naphthalene rings are named by citing a numerical prefix denoting the number of naphthalene rings followed by the term '-naphthylene'.

Examples:

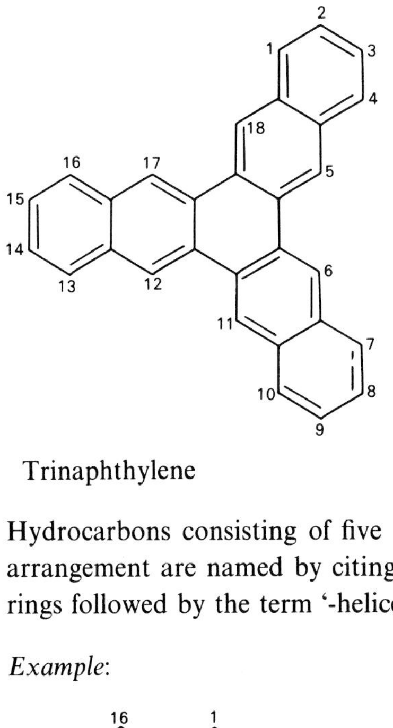

Trinaphthylene 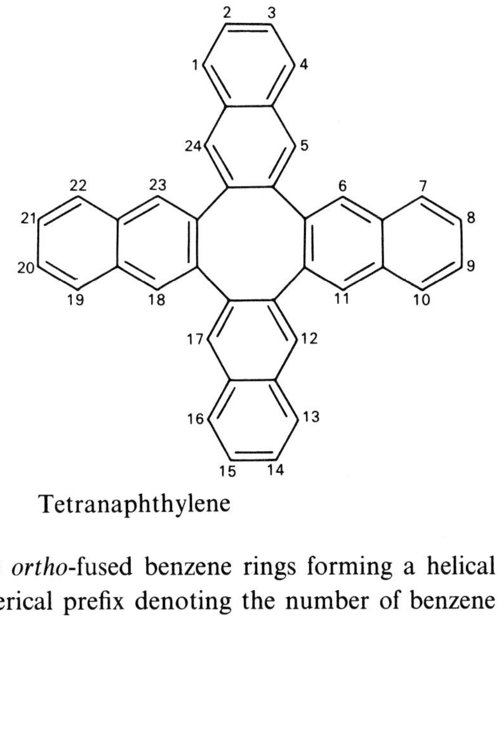   Tetranaphthylene

R-2.4.1.3.6   Hydrocarbons consisting of five or more *ortho*-fused benzene rings forming a helical arrangement are named by citing a numerical prefix denoting the number of benzene rings followed by the term '-helicene'.

Example:

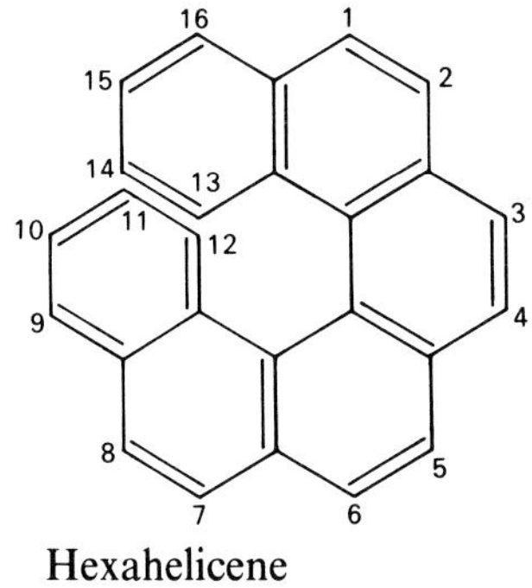

Hexahelicene

R-2.4.1.4   There are several methods that can be used to derive semisystematic names for heterocyclic components of fused ring systems.

R-2.4.1.4.1   Heterotricyclic ring systems consisting of two benzene rings *ortho*-fused to a 1,4-diheteroatomic six-membered monocyclic ring in which the heteroatoms are different are named by adding the prefix 'pheno-' to the Hantzsch–Widman name of the heteromonocycle (see R-2.3.3.3)[28].

Examples:

Phenoxathiine   10*H*-Phenoselenazine

R-2.4.1.4.2   Heterotricyclic ring systems consisting of two benzene rings *ortho*-fused to a 1,4-diheteroatomic six-membered monocyclic ring in which the heteroatoms are the same are named by adding the replacement prefix for the heteroatom (see Table 3) to the term '-anthrene', with elision of an 'a'. As exceptions, the names 'phenazine' and 'phenomercurine' are retained.

Example:

Thianthrene

R-2.4.2   Bridged parent hydrides – extension of the von Baeyer system

R-2.4.2.1   *Bicyclic ring systems.* Saturated homogeneous bicyclic systems having two or more atoms in common, are named by prefixing 'bicyclo-' to the name of the acyclic parent hydride that has the same total number of skeletal atoms; heteroatoms in an otherwise hydrocarbon system are indicated by replacement nomenclature using 'a' prefixes (see R-9.3). The number of skeletal atoms in each of the three acyclic chains (bridges) connecting the two common atoms (bridgeheads[29]) is given by arabic numbers cited in descending numerical order separated by full stops and enclosed in square brackets.

The system is numbered starting with one of the bridgeheads and proceeding through the longest bridge to the second bridgehead, continuing back to the first bridgehead by means of the longer unnumbered bridge; these two bridges constitute the 'main ring' of the system. Numbering is completed by numbering the remaining bridge (the shortest) beginning with the atom next to the first bridgehead.

[28] In names containing the prefix 'pheno-' associated with a Hantzsch–Widman name, the final 'e' is optional.
[29] A 'bridge' is a valence bond, an atom, or an unbranched chain of atoms connecting two different atoms that are already part of a cyclic system of atoms. The two skeletal atoms connected by the bridge are called 'bridgeheads'.

Examples:

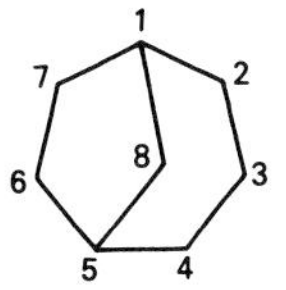

Bicyclo[3.2.1]octane

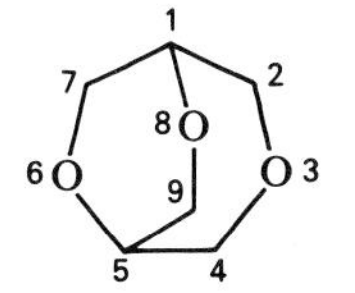

3,6,8-Trioxabicyclo[3.2.2]nonane

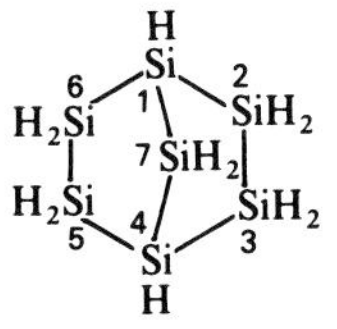

Bicyclo[2.2.1]heptasilane

This method has also been used for bicyclic systems of alternating skeletal atoms.

Example:

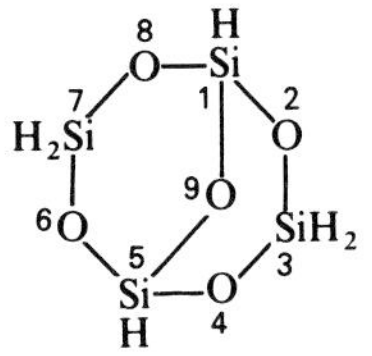

Bicyclo[3.3.1]tetrasiloxane

R-2.4.2.2   ***Polycyclic ring systems***[30]. Polycyclic analogues of saturated bicyclic ring systems (see R-2.4.2.1) are named by using the prefixes 'tricyclo-', 'tetracyclo-', etc., in place of 'bicyclo-'. The number of atoms in the additional bridges (called 'secondary bridges') is indicated by arabic numbers separated by full stops and cited, in decreasing numerical order, following those describing the largest bicyclic system. The location of each secondary bridge is indicated by the arabic number locants of the structure already numbered, which are cited as superscripts to the arabic number denoting its length and separated by a comma. When there are secondary bridges of equal length, they are cited in order of the increasing value of their lower-numbered bridgehead atom. The secondary bridges are numbered in decreasing order of size. Numbering of each bridge follows from the bridgehead already numbered proceeding from its higher numbered end. If bridges of equal lengths are present, numbering begins with the bridge having the highest numbered bridgehead atom.

---

[30] This subsection illustrates the extension of the von Baeyer system to polycyclic systems. For more details, see Rule A-32, in the IUPAC *Nomenclature of Organic Chemistry*[1].

Examples:

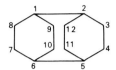

Tricyclo[4.2.2.2^{2,5}]dodecane

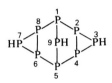

Tetracyclo[3.3.1.0^{2,4}.0^{6,8}]nonaphosphane

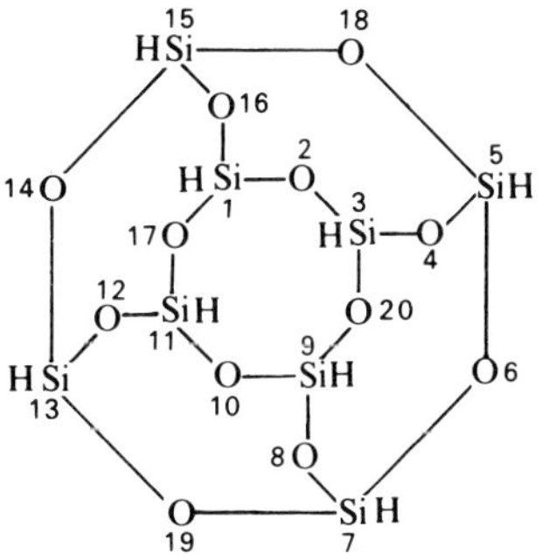

Pentacyclo[9.5.1.1^{3,9}.1^{5,15}.1^{7,13}]octasiloxane

Prefixes such as *1Si*- and *1N*-are used in polycyclic systems of alternating skeletal atoms when it is necessary to indicate the atom at the bridgehead that is to have the locant '1'.

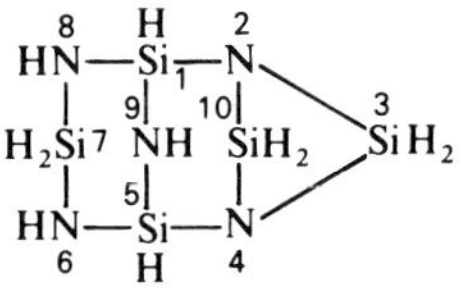

1Si-Tricyclo[3.3.1.1^{2,4}]pentasilazane

R-2.4.3   **Spiro parent hydrides**

A 'spiro union' is a linkage between two rings consisting of a single atom common to both. A 'free spiro union' is a linkage that constitutes the only union direct or indirect between the two rings[31]. The common atom is designated as the 'spiro atom'. According to the number of spiro atoms present, the compounds are distinguished as monospiro, dispiro, trispiro, etc., ring systems. The following recommendations apply only to the naming of parent hydrides containing free spiro unions[32].

R-2.4.3.1   *Monospiro parent hydrides* consisting of two homogeneous saturated monocyclic rings are named by placing 'spiro' before the name of the acyclic parent hydride with the same total number of skeletal atoms; heteroatoms, if any, in an otherwise hydrocarbon structure are designated by replacement nomenclature, i.e., by 'a' prefixes (see R-9.3)

---

[31] An example of a compound where the spiro union is not free is:

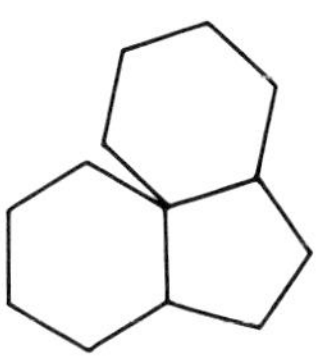

[32] In earlier editions of the IUPAC *Nomenclature of Organic Chemistry*, two methods were described for the naming of spiro parent hydrides (see Rules A-41 and 42, pp. 38–41 and Rules B-10 and 11, pp. 72–73 in the 1979 edition[1]). In these recommendations, only the first of these methods has been retained.

placed before the spiro prefix. The numbers of skeletal atoms linked to the spiro atom in each ring are indicated by arabic numbers separated by a full stop, cited in ascending order, and enclosed in square brackets; this descriptor is placed between the spiro prefix and the name of the parent hydride.

Numbering starts with a ring atom next to the spiro atom and proceeds first through the smaller ring, if one is smaller, and then through the spiro atom and around the second ring.

Examples:

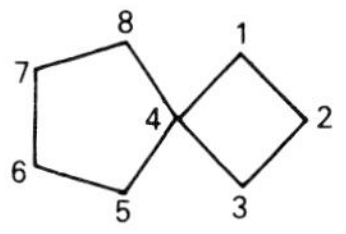

Spiro[3.4]octane 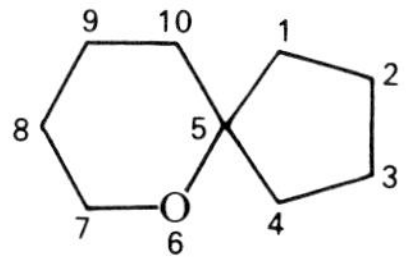   6-Oxaspiro[4.5]decane

The method is also used for monospiro systems of alternating skeletal atoms.

Example:

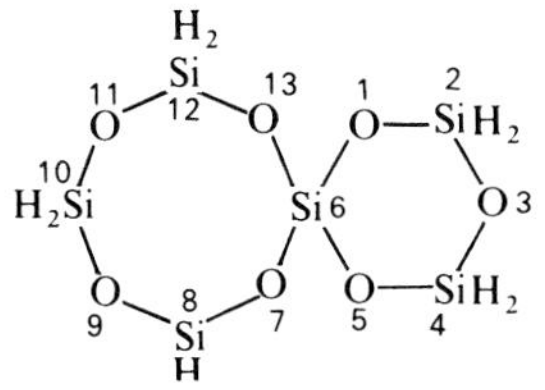

Spiro[5,7]hexasiloxane

R-2.4.3.2   ***Polyspiro parent hydrides*** consisting of unbranched assemblies of three or more saturated homogeneous monocyclic rings are named by using the prefixes 'dispiro-', 'trispiro-', 'tetraspiro-', etc., instead of 'spiro-' in front of the name of the acyclic parent hydride that has the same total number of skeletal atoms. Heteroatoms in an otherwise hydrocarbon system are designated by replacement nomenclature using 'a' prefixes (see R-9.3). The numbers of skeletal atoms linked to the spiro atom in each terminal ring and between the spiro atoms in the other rings are given by arabic numbers separated by full stops cited in the same order as the numbers proceed around the ring and enclosed in square brackets. Numbering begins with a ring atom next to the terminal spiro atom of the smaller terminal ring, proceeding around that terminal ring through its terminal spiro atom and, by the shortest path, through each of the other spiro atoms, around the other terminal ring, and then back to the first terminal ring.

Examples:

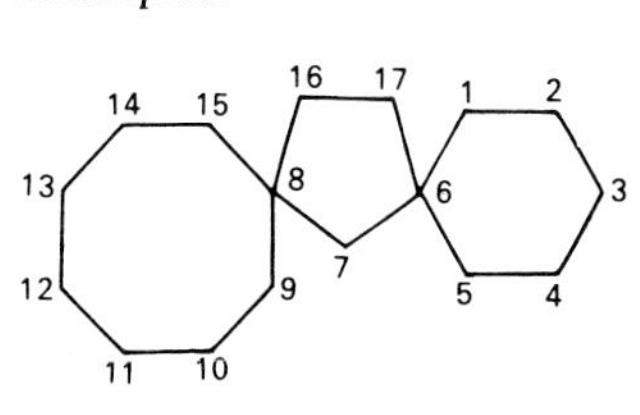

Dispiro[5.1.7.2]heptadecane 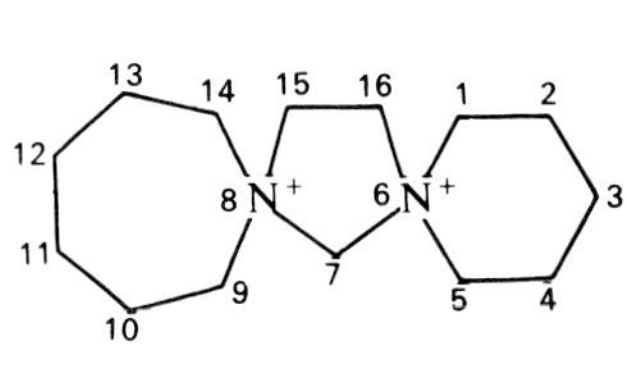   6,8-Diazoniadispiro[5.1.6.2]hexadecane dichloride

Note: Extension of this procedure to branched polyspiro systems may lead to ambiguity.

52

R-2.4.3.3   ***Spiro parent hydrides containing polycyclic ring systems,*** such as a fused ring system, are named by placing the prefix 'spiro-', 'dispiro-', 'trispiro-', etc., in front of the names of the components, which are cited in order of occurrence beginning with the terminal component earliest in alphabetical order, and enclosed in square brackets. Established numbering of each component is retained, but those of the second cited and succeeding components are primed serially.

Examples:

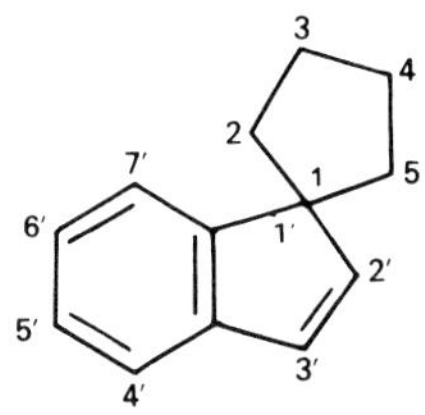

Spiro[cyclopentane-1,1'-indene] 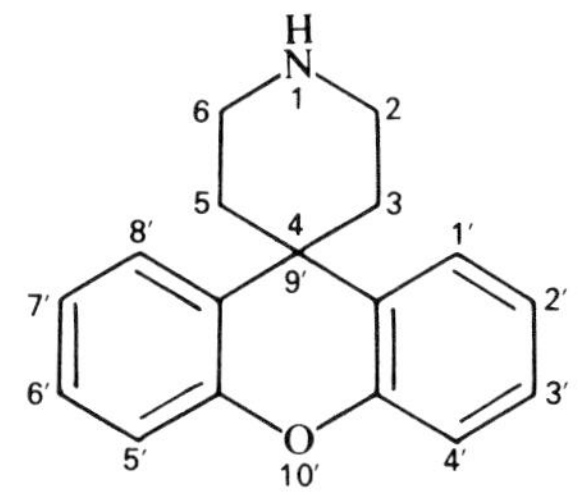   Spiro[piperidine-4,9'-xanthene]

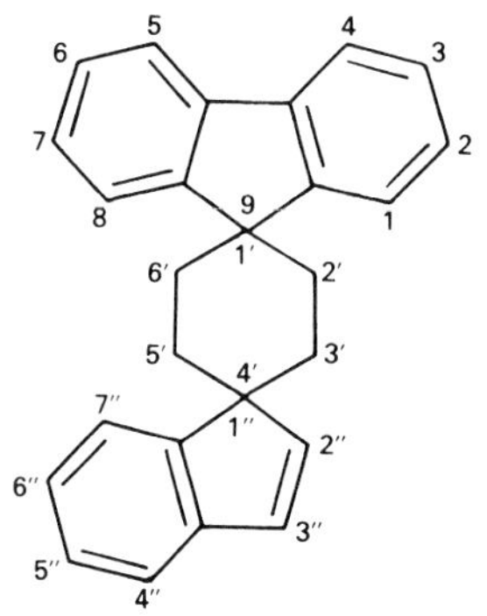

Dispiro[fluorene-9,1'-cyclohexane-4',1''-indene]

R-2.4.4   **Ring assemblies**

Two or more identical cyclic systems (whether mono- or bicyclic) directly joined to each other by double or single bonds are called 'ring assemblies' when the number of such direct ring junctions is one less than the number of cyclic systems involved.

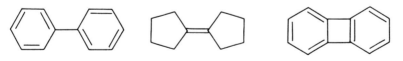

Ring assemblies   Fused polycyclic system

R-2.4.4.1   ***Assemblies of two identical cyclic systems*** are named in one of two ways: (a) by placing the prefix 'bi-' before the name of the corresponding substituent group (see R-2.5) enclosed in parentheses, if necessary (additive operation); or (b) by placing the prefix 'bi-' before the name of the corresponding parent hydride enclosed in parentheses[33], if necessary (conjunctive operation).

---

[33] Parentheses may avoid confusion with von Baeyer names.

Examples:

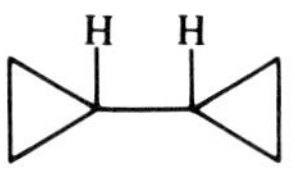

(a) 1,1'-Bi(cyclopropyl)
(b) 1,1'-Bi(cyclopropane)

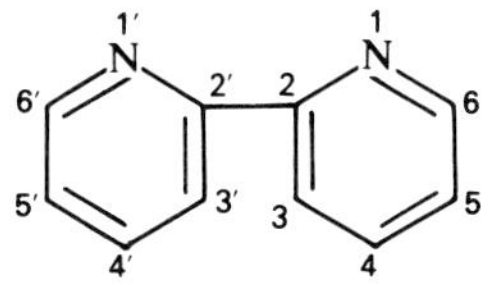

(a) 2,2'-Bipyridyl
(b) 2,2'-Bipyridine

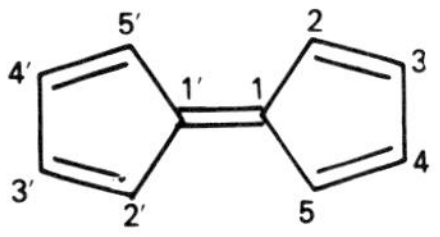

(a) 1,1'-Bi(cyclopenta-2,4-dien-1-ylidene)[34]

R-2.4.4.2   **Unbranched assemblies consisting of three or more identical ring systems** are named by placing an appropriate numerical prefix 'ter-', 'quater-', 'quinque-', etc. (see R-0.1.4.3), before the name of the parent hydride corresponding to the repetitive unit.

Examples:

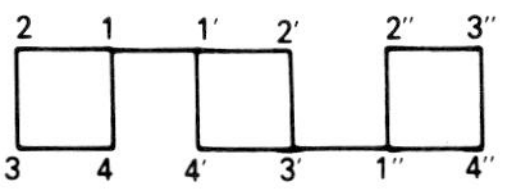

1,1':3',1''-Tercyclobutane

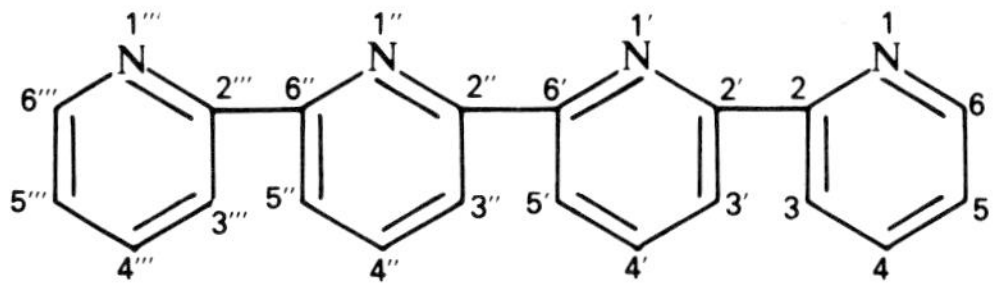

2,2':6',2'':6'',2'''-Quaterpyridine

As exceptions, unbranched assemblies consisting of benzene rings are named by using the appropriate numerical prefix with the substituent prefix name 'phenyl'.

Examples:

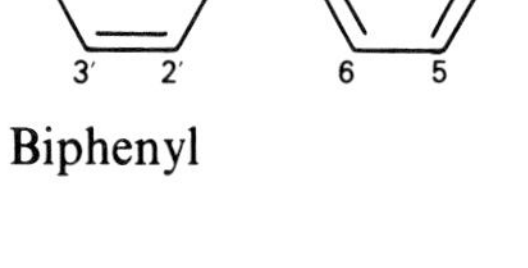

Biphenyl

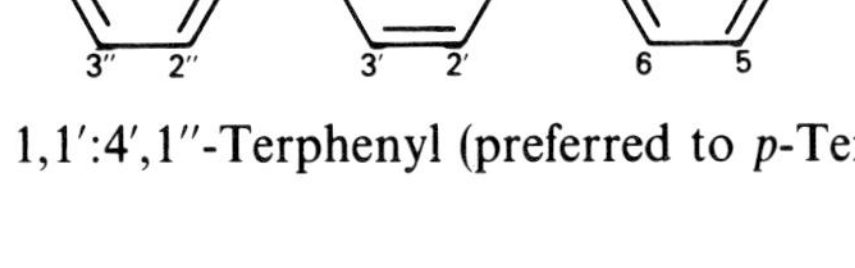

1,1':4',1''-Terphenyl (preferred to *p*-Terphenyl)

----

[34] Method (b) has also been used for naming ring assemblies connected by a double bond. The presence of a double bond was indicated by the capital Greek letter Δ and the point of attachment on the ring was given by superscript locant numbers. This method is not continued in these recommendations; accordingly, assemblies of three or more identical ring systems interconnected by double bonds cannot be named as ring assemblies.

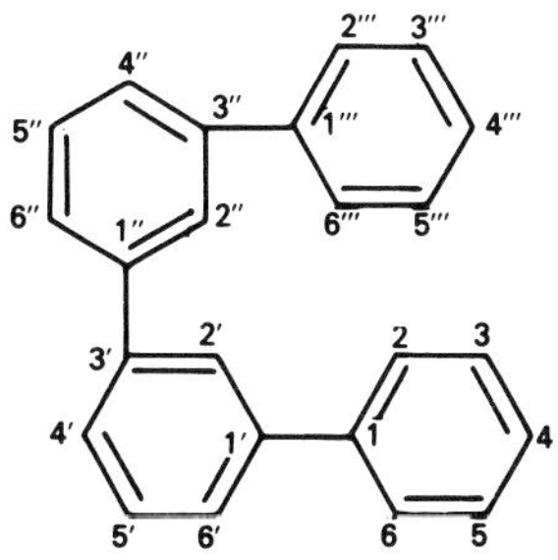

1,1':3',1":3",1'''-Quaterphenyl (preferred to *m*-Quaterphenyl)

R-2.4.5   **Cyclophanes**
The term 'cyclophanes' applies to cyclic systems consisting of ring(s) or ring system(s) having the maximum number of noncumulative double bonds connected by saturated and/or unsaturated chains. Names for cyclophanes are formed by the operation of replacing a single atom of a cyclic parent 'phane' structure by a cyclic structure.

Example[35]:

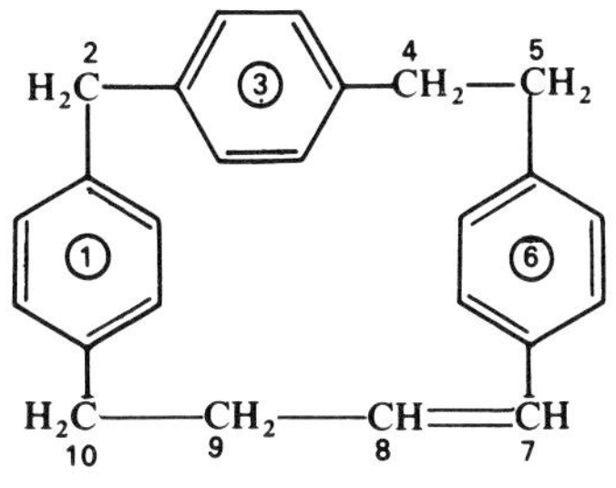

R-2.4.6   **Natural product parent hydrides**
Many naturally occurring compounds belong to well-defined structural classes, each of which can be characterized by a set of cognate parent structures that are closely related structurally. In naming natural product structures, the main objective is to choose a parent structure that includes as much as possible of the configurational detail common to the relevant class of natural products; such a parent structure has become known as a stereoparent. A numbering pattern established among a group of structurally related natural products is normally used for numbering the skeletal atoms of the stereoparent. Nomenclature for natural products is described briefly in Section F of the 1979 edition of the IUPAC *Nomenclature of Organic Chemistry*[1].

[35] Details of a nomenclature for cyclophanes and the name for this example will be given in a forthcoming publication.

Examples:

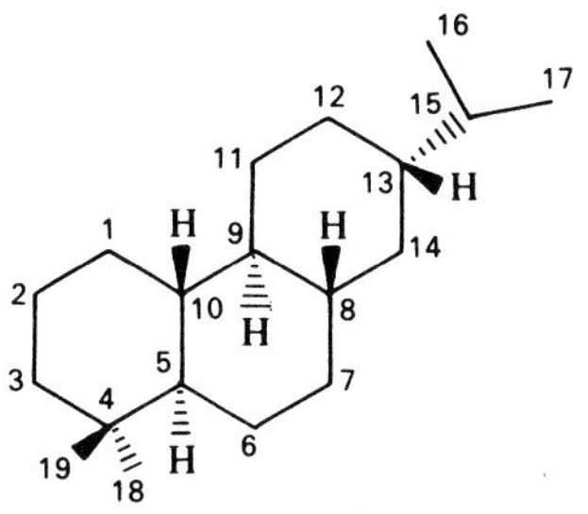

Abietane

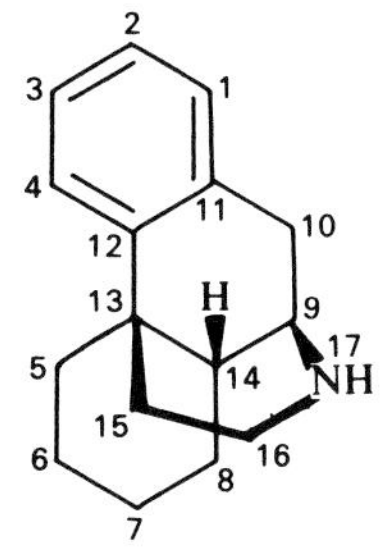

Morphinan

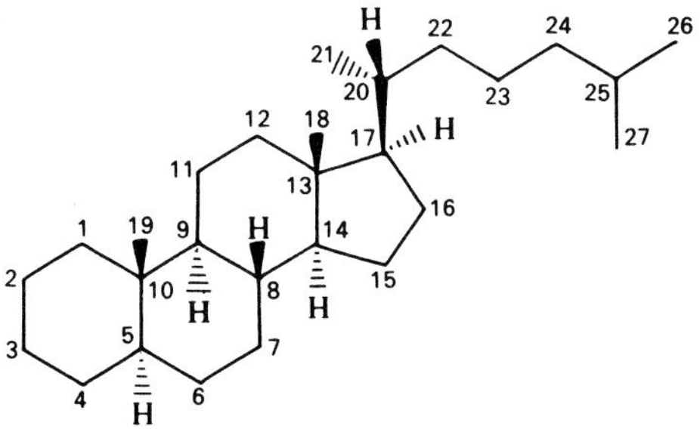

5α-Cholestane

R-2.5   SUBSTITUENT PREFIX NAMES DERIVED FROM
             PARENT HYDRIDES

The presence of one or more free valence(s) derived from the loss of one or more
hydrogen atoms from a parent hydride is denoted by suffixes such as '-yl', '-diyl',
'-ylidene', '-triyl', '-ylidyne'.

Monovalent	Divalent	Trivalent	Tetravalent	etc.
-yl	-diyl -ylidene	-triyl -ylidyne -ylylidene	-tetrayl -ylylidyne -diylidene -diylylidene	etc.

Note: In these recommendations, the suffixes '-ylidene' and '-ylidyne' are used only to
indicate the attachment of a substituent to a parent hydride or parent substituent by a
double or triple bond, respectively.

These suffixes are used according to two methods as follows:
(a) The suffixes '-yl', '-ylidene', and '-ylidyne' replace the ending '-ane' of the parent
hydride name. The atom with the free valence terminates the chain and always has the
locant '1', which is omitted from the name. This method is recommended only for
saturated acyclic and monocyclic hydrocarbon substituent groups and for the mononuclear parent hydrides of silicon, germanium, tin, lead, and boron.
(b) *More general method.* Any of these suffixes may be added to the name of the parent
hydride with elision of a terminal 'e', if present, before suffixes beginning with 'y'. The

56

atoms with free valences are given numbers as low as is consistent with any established numbering of the parent hydride[36]; except for the suffix '-ylidyne', the locant number '1' must always be cited.

Examples:

$CH_3-CH_2-CH_2-CH_2-CH_2-$
5   4   3   2   1
(a) Pentyl
(b) Pentan-1-yl

$CH_3-CH_2-CH_2-\overset{|}{C}H-CH_3$
(a) 1-Methylbutyl
(b) Pentan-2-yl

$CH_3-\overset{|}{C}H-$
2   1
(b) Ethane-1,1-diyl

$CH_3CH=$
(a) Ethylidene
(b) Ethanylidene

$CH_3-CH_2-CH_2-CH=$
4   3   2   1

(a) Butylidene
(b) Butan-1-ylidene

$CH_3-CH_2-\overset{|}{\underset{|}{C}}-$

3   2   1
(b) Propane-1,1,1-triyl

$CH_3-CH_2-C\equiv$
(a) Propylidyne
(b) Propanylidyne

$CH_3-CH_2-\overset{|}{C}=$
3   2   1
(b) Propan-1-yl-1-ylidene

$CH_3-\overset{||}{C}-CH_3$
(a) 1-Methylethylidene
    Isopropylidene (when unsubstituted)[37]
(b) Propan-2-ylidene

$CH_3-\overset{|}{\underset{|}{C}}-CH_3$

(a) 1-Methylethane-1,1-diyl
(b) Propane-2,2-diyl

$-CH_2-CH_2-CH_2-$
(b) Propane-1,3-diyl[38]

$CH_3-\overset{|}{C}H-CH_2-CH_2-$
(b) Butane-1,3-diyl

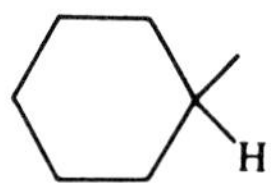

(a) Cyclohexyl
(b) Cyclohexan-1-yl

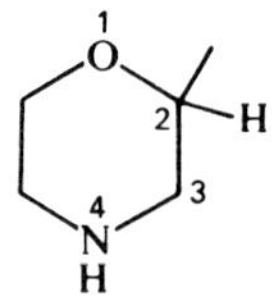

(b) Morpholin-2-yl[39]

---

[36] As an alternative, the names 'methylene', 'ethylene', and 'phenylene' may be used to designate $-CH_2-$ (methanediyl), $-CH_2-CH_2-$ (ethane-1,2-diyl), and $-C_6H_4-$ (benzenediyl), respectively. The suffix '-ylene' should not be used with the mononuclear silicon, germanium, tin, or lead parent hydrides for substituent names but is allowed for the $:CH_2$ and $:SiH_2$ radicals (see R-5.8.1.2)

[37] Used when attached to two different atoms only in special situations, such as naming of acetals of carbohydrates and related structures (see Table 19b, p. 163).

[38] Names such as trimethylene, used in the 1979 Edition of the IUPAC *Nomenclature of Organic Chemistry*[1] are acceptable for special uses, such as in polymer nomenclature.

[39] The name 'morpholino' is commonly used and is an acceptable alternative for morpholin-4-yl.

The contracted names adamantyl, naphthyl, anthryl, phenanthryl (R-9.1, Table 22, p. 166), and furyl, pyridyl, isoquinolyl, quinolyl, piperidyl[40] (R-9.1, Table 25, p. 172) are retained. The trivial names vinyl, allyl, and phenyl (R-9.1, Table 19(b), p. 163, and thienyl, furfuryl, and thenyl (see R-9.1, Table 25, p. 172) are retained for use with no limitation on substitution; other trivial names are retained but only with limited or no substitution (see R-9.1, Table 19(b), p. 163).

---

[40] The name 'piperidino' is commonly used and is an acceptable alternative for 1-piperidyl.

R-3   Characteristic (Functional) Groups

R-3.0   INTRODUCTION

The prefixes and/or suffixes attached to a parent name specifying a particular molecular structure usually represent *substituents* of various types, which are considered to replace hydrogen atoms of a parent hydride or parent structure. It has been customary to regard such substituents as *characteristic* (or *functional*) when the link between substituent and parent is not a carbon–carbon bond, for example, –OH, =O, and –NH$_2$, but many exceptions are recognized, such as –COOH and –C≡N. It seems appropriate at this time to retain the general view of *functionality* as implying the presence of heteroatoms and/or unsaturation, but it would not be helpful to attempt to define precisely the limits of application of the term.

Carbon–carbon unsaturation in acyclic species is regarded as a special type of functionality and it is therefore treated here in Section R-3 rather than in Section R-2 (Parent Hydrides); however, its presence here (and that of hydrogenation of parent hydrides containing the maximum number of noncumulative double bonds) is anomalous in that for some purposes, for example, choice of parent, it can be regarded as part of the parent; but for others, such as numbering, it is treated like a substituent.

Section R-3 also deals with *functional parents*, i.e., structures which are treated as parent structures, having substitutable hydrogen atoms, but which possess the characteristics normally associated with functionality [e.g., phosphonic acid HP(O)(OH)$_2$].

Although, strictly speaking, ions and radicals do not fall within the concept of functionality, as described above, an ionic centre or a radical centre is treated like a function, and this treatment is also included, therefore, here in Section R-3.

R-3.1   UNSATURATION

As noted above (R-3.0), the treatment of unsaturation (of a saturated parent hydride) or saturation (of an unsaturated parent hydride) is different from that of other functions. This difference has implications for choice of parent hydride and for numbering. The following three subsections describe ways to indicate unsaturation/saturation.

R-3.1.1   **Suffixes denoting multiple bonds**

The presence of one or more double or triple bonds in an otherwise saturated parent hydride (except for parent hydrides with Hantzsch–Widman names) is denoted by changing the '-ane' ending of the name of a saturated parent hydride to one of the following:

	One	Two	Three	etc.
Double bond	-ene	-adiene	-atriene	etc.
Triple bond	-yne	-adiyne	-atriyne	etc.

The presence of both double and triple bonds is similarly denoted by endings such as '-enyne', '-adienyne', '-enediyne', etc. Numbers as low as possible are given to double and triple bonds as a set, even though this may at times give '-yne' a lower number than '-ene'. If a choice remains, preference for low locants is given to the double bonds. Only the lower locant for a multiple bond is cited except when the numerical difference between the two locants is greater than one, in which case the higher-numbered locant is cited in parentheses (see R-0.1.4).

Examples:

$CH_3–CH_2–CH_2–CH=CH–CH_3$
6   5   4   3   2   1

Hex-2-ene

$CH_3–CH=CH–CH=CH–CH_3$
6   5   4   3   2   1

Hexa-2,4-diene

$CH≡C–CH=CH–CH=CH_2$
6   5   4   3   2   1

Hexa-1,3-dien-5-yne

$NH_2–NH–N=N–NH_2$
5   4   3   2   1

Pentaaz-2-ene

$PH_2–N=P–N=PH$
5   4   3   2   1

Triphosphaza-1,3-diene

$CH_3–CH_2–C≡C–CH_3$
5   4   3   2   1

Pent-2-yne

$CH≡C–CH_2–CH=CH_2$
5   4   3   2   1

Pent-1-en-4-yne

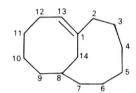

Bicyclo[6.5.1]tetradec-1(13)-ene

Use of these endings is further illustrated in Sections R-4 and R-5 (see also the 1979 edition of the IUPAC *Nomenclature of Organic Chemistry*[1], Rule A-3, p. 11)

R-3.1.2   **Hydro prefixes**

If the name of the parent hydride implies the presence of the maximum number of noncumulative double bonds (see R-2.4.1), other states of hydrogenation can usually be indicated by use of the prefix 'hydro-' together with an appropriate numerical prefix signifying the addition of hydrogen atoms. This operation is regarded as the reduction of double bonds; thus, hydrogen atoms can only be added in pairs (by use of 'dihydro-', 'tetrahydro-', etc.). 'Indicated hydrogen' (see R-1.3) is included if required by the parent hydride.

Examples:

1,4-Dihydronaphthalene

2,3,4,5-Tetrahydroazocine   6,7-Dihydro-5H-benzo[7]annulene[41]

R-3.1.3   **Dehydro prefixes**

The presence of a double bond not implied in the parent hydride name can be indicated by a 'didehydro-' prefix signifying the removal of a pair of hydrogen atoms.

Examples:

Phosphepane   1,2-Didehydrophosphepane[42]

Cholesterol   7,8-Didehydrocholesterol[43]

Similarly, the conversion of a double into a triple bond can be indicated by 'didehydro-'.

Example:

1,2-Didehydrobenzene
(formerly called 'Benzyne')

(see also 7,8-Didehydro-ε,ε-carotene under R-1.2.5.1).

[41] The traditional names obtained on the basis of Rules A-21.4 and A-23.5 of IUPAC *Nomenclature of Organic Chemistry*[1], '6,7-Dihydro-5H-benzocycloheptene' and 'Benzocyclohepta-1,3-diene', could lead to ambiguity. They are no longer encouraged. For an explanation of the use of 'annulene' as the base component in these names, see section R-2.4.1.1.
[42] However, the name '4,5,6,7-tetrahydro-3H-phosphepine' has been used traditionally and is still to be preferred in these recommendations.
[43] However, the name 'cholesta-5,7-dien-3β-ol', is preferred in IUPAC–IUBMB steroid nomenclature.

R-3.1.4 **Substituent prefix names for unsaturated/saturated parent hydrides**

When derived according to R-3.1.1–3.1.3 above, these are formed by replacing the final 'e' of the name of the unsaturated/saturated structure by an appropriate suffix given in R-2.5.

Free valence positions are preferred to positions of unsaturation denoted by the suffixes '-ene', etc., for low locant numbers[44]. As with substituent prefixes derived from saturated parent hydrides (see R-2.5), the free valences can be at any position on the parent structure. Accordingly, for acyclic unsaturated parent structures, all locants for the free valences, including '1', must be cited.

Examples:

$$CH_2 = CH-CH_2-CH_2-$$
$$\quad 4 \qquad 3 \qquad 2 \qquad 1$$

But-3-en-1-yl

$$CH_2 = CH-\overset{|}{C}H-CH_2-\overset{|}{C}H_2-CH=CH_2$$
$$\quad 7 \qquad 6 \qquad 5 \qquad 4 \qquad 3 \qquad 2 \qquad 1$$

Hepta-1,6-diene-3,5-diyl

Cyclopent-3-ene-1,2-diyl

R-3.2 SPECIFICATION OF CHARACTERISTIC GROUPS

R-3.2.1 Prefixes and suffixes

R-3.2.1.1 The presence of a characteristic group can be denoted by a prefix or suffix attached to the parent name (with elision of terminal 'e', if present). Such prefixes and suffixes are illustrated in Table 5. Additional details for the use of functional prefixes and suffixes are found in Section R-4 (Name Construction).

R-3.2.1.2 Affixes used to denote ionic or radical centres in a parent structure, classified according to the type of formal operation by which the ion or radical may be considered to be derived (see R-5.8), are given in Table 6. Suffixes are added to the name of a parent hydride in the customary manner or serve as endings to names of characteristic groups. When needed, ionic centres are expressed as prefixes that terminate with endings given in the column labelled 'Substituent prefix ending', for example, ethan-1-ylium-1-yl, pyridinio, sulfonato, and propan-2-id-2-yl; for radical centres, the prefix 'ylo-' is added in front of the name of the substituent prefix, for example, 2-yloethyl. In replacement nomenclature, the replacement prefixes are modified to end as shown in the column labelled 'Replacement prefix endings', for example, azonia and boranylia.

Two or more such operations are indicated by an appropriate multiplying affix, for example, '-diide', and '-tris(ylium)'. Additional details for the use of ionic prefixes and suffixes are given in Section R-5.8 and in a separate publication[45].

---

[44] Relative preference of free valence positions *vs.* 'hydro' and 'dehydro' positions is under study by the Commission.

[45] International Union of Pure and Applied Chemistry. Organic Chemistry Division. Commission on Nomenclature of Organic Chemistry, 'Revised Nomenclature of Radicals, Ions, Radical Ions and Related Species'. *Pure Appl. Chem.* **65**, 1357–1455 (1993).

Table 5 Suffixes and prefixes for some important characteristic groups in substitutive nomenclature (in alphabetical order[a])

Class	Formula[b]	Prefix	Suffix
Acid halides	$-CO$–halogen $-(C)O$-halogen	halocarbonyl-[c] —	-carbonyl halide -oyl halide
Alcoholates, Phenolates	$-O^-$	oxido-	-olate
Alcohols, Phenols	$-OH$	hydroxy-	-ol
Aldehydes	$-CHO$ $-(C)HO$	formyl- oxo-	-carbaldehyde -al
Amides	$-CO-NH_2$ $-(C)O-NH_2$	carbamoyl- —	-carboxamide -amide
Amidines	$-C(=NH)-NH_2$ $-(C)(=NH)-NH_2$	carbamimidoyl-[d] —	-carboximidamide[d] -imidamide[d]
Amines	$-NH_2$	amino-	-amine
Carboxylates	$-COO^-$ $-(C)OO^-$	carboxylato- —	-carboxylate -oate
Carboxylic acids	$-COOH$ $-(C)OOH$	carboxy- —	-carboxylic acid -oic acid
Ethers	$-OR^e$	(R)-oxy-	—
Esters (of carboxylic acids)	$-COOR^e$ $-(C)OOR^e$	(R)-oxycarbonyl- —	(R) . . . carboxylate (R) . . . oate
Hydroperoxides	$-O-OH$	hydroperoxy-	—
Imines	$=NH$ $=NR$	imino- (R)-imino-	-imine
Ketones	$>(C)=O$	oxo-	-one
Nitriles	$-C\equiv N$ $-(C)\equiv N$	cyano- —	-carbonitrile -nitrile
Peroxides	$-O-OR^e$	(R)-peroxy-	—
Salts (of carboxylic acids)	$-COO^- M^+$ $-(C)OO^- M^+$	— —	(cation) . . . carboxylate (cation) . . . oate
Sulfides	$-SR^e$	(R)-sulfanyl-	—
Sulfonates	$-SO_2-O^-$	sulfonato-	-sulfonate
Sulfonic acids	$-SO_2-OH$	sulfo-	-sulfonic acid
Thiolates	$-S^-$	sulfido-	-thiolate
Thiols	$-SH$	sulfanyl-[f]	-thiol

[a] For a suggested priority of classes for choice as principal characteristic group, see R-4.1, Table 10.

[b] (C) designates a carbon atom included in the name of the parent hydride and does not belong to a group designated by a suffix or a prefix.

[c] Halocarbonyl replaces haloformyl which was used in the 1979 edition of the IUPAC *Nomenclature of Organic Chemistry*[1].

[d] The suffixes '-amidine' and '-carboxamidine', and the prefix 'amidino-' were used in previous editions of the IUPAC *Nomenclature of Organic Chemistry*[1].

[e] R designates a substituent group derived from a parent hydride by loss of a hydrogen atom.

[f] In these recommendations, the prefix 'sulfanyl-' is preferred to 'mercapto-' which was used in previous editions of the IUPAC *Nomenclature of Organic Chemistry*[1].

Table 6 Affixes for ionic and radical centres in parent structures

	Operation	Suffix	Substituent prefix ending	Replacement prefix ending
Radicals (see R-5.8.1)	Loss of H·	-yl	—[a]	—
Cations (see R-5.8.2)	Loss of H⁻	-ylium	-yliumyl	-ylia
	Addition of H⁺	-ium	-io -iumyl	—
		-onium	-onio	-onia
Anions (see R-5.8.3)	Loss of H⁺	-ide	-idyl	-ida
		-ite	—	—
		-ate	-ato	—
	Addition of H⁻	-uide	-uidyl	-uida[b]

[a] '-ylo' is not a substituent prefix ending but a prefix to another prefix, e.g., ylomethyl-.
[b] In the provisional Section D of the 1979 edition of the IUPAC *Nomenclature of Organic Chemistry*[1], the ending '-ata' was proposed.

R-3.2.2   **Functional modifiers**

Many derivatives of principal characteristic groups or functional parent compounds (see R-3.3) may be named by name modifiers which consist of one or more separate words placed after the name of the parent structure. This method is most useful in an indexing environment.

Note: For indexing purposes, this method is used in *Beilstein* for esters, acyl halides, amides, hydrazides, lactones, lactams and nitrogen derivatives of carbonyl compounds (oximes, hydrazones, etc.). For indexing purposes, this method is also used in Chemical Abstracts index nomenclature for anhydrides, esters, hydrazides, hydrazones and oximes.

Examples:

CH_3-CH_2-CHO
Propanal

$CH_3-CH_2-CH(OCH_3)_2$
Propanal dimethyl acetal

$CH_3-CH_2-CH=NOH$
Propanal oxime

$CH_3-CH_2-CH=NNH_2$
Propanal hydrazone

CH_3-CH_2-COOH
Propionic acid

$CH_3-CH_2-COO-CH_3$
Propionic acid methyl ester

$CH_3-CH_2-CO-NH-NH_2$
Propionic acid hydrazide

Note: The use, in this section and elsewhere in this guide, of systematic names, such as *propanal* and *propanoic acid*, rather than retained names (see R-9.1), such as propionaldehyde and propionic acid, illustrates that, even though some names are retained, this does not preclude use of the more systematic names.

R-3.3 FUNCTIONAL PARENT COMPOUNDS AND DERIVED SUBSTITUENT GROUPS

Structures which are treated as parent compounds in substitutive nomenclature although they have characteristics normally associated with functionality are illustrated in Table 7.

Table 7 Functional parent acids and derived substituent groups of nitrogen and phosphorus[46]

Parent acids		Substituent groups	Prefixes
$NH(O)(OH)_2$	Azonic acid	$-N(O)(OH)_2$	azono- dihydroxynitroryl-
		$> NH(O)$	azonoyl- hydronitroryl-
$NH_2(O)(OH)$	Azinic acid	$-NH(O)(OH)$	hydrohydroxynitroryl-
		$-NH_2(O)$	azinoyl- dihydronitroryl-
		$> N(O)(OH)$	hydroxynitroryl-
		$-\overset{\mid}{\underset{\mid}{N}}(O)$	nitroryl-
$PH(OH)_2$	Phosphonous acid	$-P(OH)_2$ $-P(OH)O^-$	dihydroxyphosphanyl- hydroxyoxidophosphanyl-
$PH_2(OH)$	Phosphinous acid	$-PH(OH)$	hydroxyphosphanyl-
		$> P(OH)$	hydroxyphosphanediyl-
		$= P(OH)$	hydroxyphosphanylidene-
		$-P(OCH_3)_2$	dimethoxyphosphanyl-
		$> P(O)(O^-)$	phosphinato-
$PH(O)(OH)_2$	Phosphonic acid	$-P(O)(OH)_2$	phosphono- dihydroxyphosphoryl-
		$> PH(O)$	phosphonoyl- hydrophosphoryl-
		$-P(O)(O^-)_2$	phosphonato-
		$-P(O)(OH)(O^-)$	hydroxyoxidophosphoryl-
$PH_2(O)(OH)$	Phosphinic acid	$-PH(O)(OH)$	hydrohydroxyphosphoryl-
		$-PH_2(O)$	phosphinoyl- dihydrophosphoryl-
		$> P(O)(OH)$	hydroxyphosphoryl-
		$-P(O)(OH)(OCH_3)$	hydroxymethoxyphosphoryl-
		$-\overset{\mid}{\underset{\mid}{P}}(O)$	phosphoryl-

[46] Arsenic and antimony functional parent acids and derived substituent groups may be named like the analogous phosphorus acids using the stems 'ars-' and 'stib-' in place of 'phosph-'.

R-3.4   FUNCTIONAL REPLACEMENT
             The replacement of oxygen atoms or hydroxy groups by other atoms or groups can be
             described by prefixes attached to, or by infixes inserted into, names of characteristic
             groups, functional parent compounds or trivial names. For example, 'thio' indicates
             replacement of oxygen by sulfur in the suffixes '-thiosulfonic acid' and '-carbothioic acid',
             and in the functional parent name 'phosphorothioic acid'. Similarly, 'peroxo' or 'peroxy'
             indicates replacement of an oxygen atom by the –O–O– group in the suffix '-peroxycar-
             boxylic acid', in the functional parent name 'phosphoroperoxoic acid', and in the trivial
             name 'peroxyacetic acid'. A list of prefixes and/or infixes is given in Table 8.

Examples:

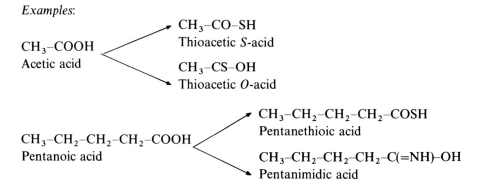

Table 8 Functional replacement prefixes and infixes

Prefix	Infix[a]	Replaced atom(s)	Replacing atom(s)
Amido-	-amido-	–OH	–NH$_2$
Azido-	-azido-	–OH	–N$_3$
Bromo-	-bromido-	–OH	–Br
Chloro-	-chlorido-	–OH	–Cl
Cyanato-	-cyanatido-	–OH	–OCN
Cyano-	-cyanido-	–OH	–CN
Dithioperoxy-[b]	-dithioperoxo-[b,c]	–O–	–S–S–
Fluoro-	-fluorido-	–OH	–F
Iodo-	-iodido-	–OH	–I
Isocyanato-	-isocyanatido-	–OH	–NCO
Isocyano-	-isocyanido-	–OH	–CN
Nitrido-	-nitrido-	=O and –OH	$\equiv$N $\begin{cases} \text{or} -N< \\ \text{or} =N- \end{cases}$
Thiocyanato-[b]	-thiocyanatido-[b,c]	–OH	–SCN
Isothiocyanato-[b]	-isothiocyanatido-[b,c]	–OH	–NCS
Imido-	-imido-	=O	=NH
Hydrazido-	-hydrazido-	–OH	–NH–NH$_2$
Peroxy-	-peroxo-	–O–	–O–O–
Seleno-	-seleno-	=O or –O–	=Se or –Se–
Telluro-	-telluro-	=O or –O–	=Te or –Te–
Thio-	-thio-	=O or –O–	=S or –S–
Thioperoxy-[b]	-thioperoxo-[b,c]	–O–	–OS– or –SO–

[a] The final 'o' of the infix is generally elided when followed by the vowels 'a', 'i', or 'o'.
[b] Selenium and tellurium analogues are named by using the replacement affixes 'seleno' and 'telluro'
in place of 'thio', e.g., selenoperoxy- and -selenoperoxo-.
[c] The infixes should always be enclosed in parentheses in a name of a structure to avoid the
possibility of misinterpretation.

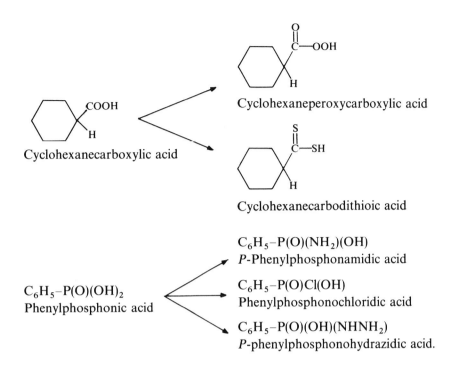

Cyclohexanecarboxylic acid

Cyclohexaneperoxycarboxylic acid

Cyclohexanecarbodithioic acid

$C_6H_5-P(O)(OH)_2$
Phenylphosphonic acid

$C_6H_5-P(O)(NH_2)(OH)$
P-Phenylphosphonamidic acid

$C_6H_5-P(O)Cl(OH)$
Phenylphosphonochloridic acid

$C_6H_5-P(O)(OH)(NHNH_2)$
P-phenylphosphonohydrazidic acid.

R-4 Guide to Name Construction

R-4.0   INTRODUCTION

In this Section, principles regarding name construction are presented. It is recognized that in chemical discussions it may sometimes be convenient to depart from rigorous rules in order to provide a name more appropriate to the chemical intent or to avoid obscuring an important feature. However, such deviations should not be allowed without good reason, and names so derived are not recommended for general use.

The application of the general principles discussed in Section R-4 will not necessarily lead to a unique name but the name obtained should be unambiguous. Recommendations leading to a unique preferred name when followed closely are in preparation and will be contained in a separate publication. The latter goal is not necessarily the aim of a practising chemist who wants to communicate with his colleagues in familiar, well understood terms, but it may be appropriate to those who cite chemical names in legislative documents, or for indexing and retrieval.

R-4.1   GENERAL PRINCIPLES

The formation of the systematic name for an organic compound involves several steps, to be taken as far as they are applicable[47] in the following order:

(a) from the nature of the compound, determine the type(s) of nomenclature operations (see Section R-1.2) to be used. Although the so-called 'substitutive nomenclature' is emphasized in these recommendations, other kinds of names, for example, functional class names, are often given, usually as alternatives;

(b) determine the kind of characteristic group to be cited as suffix (if any) or as a functional class name. Only one kind of characteristic group (known as the principal group) can be cited as suffix or functional class name[48]. All substituents not so cited must be specified as prefixes;

(c) determine the parent hydride, including any appropriate nondetachable prefixes [detailed rules for choice of the principal chain, the preferred ring or ring system, the functional parent compound, or conjunctive components are described in the 1979 edition of the IUPAC *Nomenclature of Organic Chemistry*[1] (see, for example, Rule C-12, p. 92)];

---

[47] Occasionally, an author may wish to convey a particular emphasis by assigning a name which departs from standard priority considerations. So long as this is adequately explained and appropriate consequences (e.g., for numbering) accepted and logically treated so as to preserve freedom from error and ambiguity, this set of procedures may still be applicable. However, names so generated are not recommended for general use.

[48] This restriction does not apply to suffixes denoting unsaturation, such as '-ene' and '-yne' (see R-3.1), which are not regarded as principal characteristic groups; nor to the ionic and radical suffixes, such as '-ium', '-ide', and '-yl' (see R-3.2.1.2), which may be combined and may also be added to certain principal characteristic group suffixes, for example, ethanaminium and pyridin-1-ium-3-yl.

(d) name the parent hydride and the principal characteristic group, if any, or the functional parent compound;

(e) determine infixes and/or prefixes [with the appropriate multiplying prefixes (see Table 11)], and number the structure as far as possible[49];

(f) name the detachable substitutive prefixes and complete the numbering of the structure, if necessary;

(g) assemble the components into a complete name, using alphabetical order for all substitutive prefixes.

In substitutive nomenclature, some characteristic groups can be denoted either as prefixes or suffixes (see Table 5), but others only as prefixes (see Table 9). Functional class names differ in that a separate word (or suffix in some languages) designating the name of a functional class is associated with a 'radical' name designating the remainder of the structure.

Table 9 Characteristic groups cited only as prefixes in substitutive nomenclature (non-limiting list)

Characteristic group	Prefix
$-Br$	Bromo-
$-Cl$	Chloro-
$-ClO$	Chlorosyl-
$-ClO_2$	Chloryl-
$-ClO_3$	Perchloryl-
$-F$	Fluoro-
$-I$	Iodo-
$-IO$	Iodosyl-
$-IO_2$	Iodyl- (replaces iodoxy-)
$-I(OH)_2$	Dihydroxy-λ^3-iodanyl- (replaces dihydroxyiodo-)
$-IX_2$[a]	Dihalo-λ^3-iodanyl- (replaces dihaloiodo-)
$=N_2$	Diazo-
$-N_3$	Azido-
$-NO$	Nitroso-
$-NO_2$	Nitro-
$-OR$[b]	(R)-oxy-
$-SR$[b]	(R)-sulfanyl- (and similarly, (R)-selanyl- and (R)-tellanyl-)
$-SH_3$	λ^4-Sulfanyl-

[a] X traditionally refers to a halogen atom.
[b] R designates a substituent group derived from a parent hydride by loss of a hydrogen atom.

[49] In these recommendations, hydro and dehydro prefixes are presented as nondetachable. For indexing purposes, Chemical Abstracts Service and *Beilstein* treat hydro prefixes as detachable; the former alphabetizes them with substitutive prefixes and the latter treats them as a separate kind of prefix cited after the substitutive prefixes. The whole question of order of precedence of hydro prefixes is under consideration by the Commission.

The characteristic groups that can be cited as suffixes in substitutive nomenclature are not necessarily identical with the groups designated by the name of a corresponding functional class when functional class names are formed (e.g., butanone and ethyl methyl ketone, where -one denotes =O and ketone denotes –CO–).

The characteristic groups listed in Table 9 are always cited as prefixes to the name of the parent structure (see Sections R-2 and R-3). Multiplying prefixes and locants are added as necessary.

Example:

1,2-Dichlorocyclohexane (the complete prefix is 1,2-dichloro-)

Characteristic groups other than those given in Table 9 may be cited as either suffixes, if available, or prefixes to the name of the parent hydride.

If characteristic groups other than those given in Table 9 are present, one (and only one) kind may be cited as suffix (the principal characteristic group) (but see footnote 48).

When a compound contains more than one kind of characteristic group not given in Table 9, the principal characteristic group is the one that characterizes the class occurring earliest (i.e., nearest to the top) in Table 10. All other characteristic groups are cited as prefixes. Some suffixes and prefixes to be used with the general classes listed in Table 10 are given in Table 5.

If, and only if, the complete suffix (that is, the suffix itself plus its multiplying affix, if any) begins with a vowel, a terminal 'e' (if any) of the preceding parent name is elided (see Section R-0). Elision or retention of the terminal 'e' is independent of the presence of numerals between it and the following letter (see Section R-0).

Table 10 General classes of compounds in decreasing order of priority for choosing and naming a principal characteristic group[50]

1	Radicals	13	Aldehydes followed by Thioaldehydes, Seleno-aldehydes, and Telluroaldehydes
2	Anions		
3	Cations	14	Ketones followed by Thioketones, Seleno-ketones, and Telluroketones
4	Zwitterionic compounds		
5	Acids (in the order COOH, C(O)O₂H; then their S and Se derivatives followed by sulfonic, sulfinic, selenonic, etc., phosphonic, arsonic, etc., acids)	15	Alcohols and Phenols followed by Thiols, Selenols, and Tellurols
		16	Hydroperoxides followed by Thiohydroper-oxides, Selenohydroperoxides, and Tellurohy-droperoxides
6	Anhydrides		
7	Esters	17	Amines
8	Acid halides	18	Imines
9	Amides	19	Hydrazines, Phosphanes, etc.
10	Hydrazides	20	Ethers followed by Sulfides, Selenides, and Tellurides
11	Imides		
12	Nitriles	21	Peroxides followed by Disulfides, Diselenides, and Ditellurides

[50]A detailed precedence order will be described in a forthcoming publication.

Table 11 Basic numerical terms (multiplying affixes)[51]

Number	Numerical term	Number	Numerical term	Number	Numerical term	Number	Numerical term
1	mono-[a]	11	undeca-	100	hecta-	1000	kilia-
2	di-[a]	12	dodeca-[b]	200	dicta-	2000	dilia-
3	tri-	20	icosa-[c]	300	tricta-	3000	trilia-
4	tetra-	30	triaconta-	400	tetracta-	4000	tetralia-
5	penta-	40	tetraconta-	500	pentacta-	5000	pentalia-
6	hexa-	50	pentaconta-	600	hexacta-	6000	hexalia-
7	hepta-	60	hexaconta-	700	heptacta-	7000	heptalia-
8	octa-	70	heptaconta-	800	octacta-	8000	octalia-
9	nona-	80	octaconta-	900	nonacta-	9000	nonalia-
10	deca-	90	nonaconta-				

[a] When alone, the numerical term for the number 1 is 'mono-' and that for 2 is 'di-'. In association with other numerical terms, the number 1 is represented by 'hen-' (except in the case of 'undeca-') and the number 2 by 'do-' (except in the cases of 'dicta-' and 'dilia-'). The numerical term for the number 11 is 'undeca-'. Compare *mono*chloro with *hen*icosane for the number 21 and *di*chloro with *do*cosane for the number 22.
[b] After 'dodeca-' (12), composite numerical terms are formed systematically by citing the basic terms in the order opposite to that of the constituent arabic numerals. For example, the numerical term for the number 468 is 'octahexacontatetracta-'.
[c] Chemical Abstracts Service and *Beilstein* use eicosa instead of icosa.

The basic numerical terms (multiplying affixes) given in Table 11 are used by direct joining without hyphen.

Numerical terms for use as multiplying prefixes for complex entities, such as substituted substituents, are obtained by adding the ending '-kis' to the numerical term defined as above. However, the ending '-kis' is not used with 'mono-'. As exceptions, 'bis-' is used for the number 2 and 'tris-' for 3.

Examples:

4	tetrakis-
231	hentriacontadictakis-

The following multiplying prefixes are used in names for unbranched assemblies of three or more identical repeating units.

2	bi-	5	quinque-	8	octi-
3	ter-	6	sexi-	9	novi-
4	quater-	7	septi-	10	deci-

When a substituent is itself substituted (a complex substituent, see R-4.2.4), all the subsidiary substituents are named as prefixes. The substituent bearing the subsidiary substituents is regarded as a 'parent substituent' (analogous to a parent hydride). The nomenclature of the whole substituent is subject to all the procedures adopted for

[51] International Union of Pure and Applied Chemistry. Organic Chemistry Division. Commission on Nomenclature of Organic Chemistry, 'Extension of Rules A-1.1 and A-2.5 Concerning Numerical Terms Used in Organic Nomenclature (Recommendations 1986)', *Pure Appl. Chem.*, **58**, 1693–1696 (1986).

compounds, with two exceptions, namely:

(a) that no characteristic group is expressed as a suffix (instead, a suffix such as '-yl' or '-ylidene', etc., is used); and
(b) that the point of attachment of the complex substituent has the lowest permissible locant.

When the parent hydride (principal chain, ring system), principal group and other substituents have been selected and named, the numbering of the complete compound is allocated.

Insofar as the preceding rules leave a choice, the starting point and direction of numbering of a compound are chosen so as to give lowest locants to the following structural features (if present) considered successively in the order listed until a decision is reached.

(a) fixed numbering (naphthalene, etc.)
(b) heteroatoms in heterocycles
(c) indicated hydrogen (see R-1.3)
(d) principal group named as suffix
(e) heteroatoms in an acyclic parent structure
(f) unsaturation (ene/yne)[52]
(g) substituents named as prefixes.

The various components having been selected, named and numbered, any necessary additive and subtractive modifications are made, and the complete name is assembled, prefixes being arranged in alphabetical order.

R-4.2   *Examples*:

R-4.2.1

	CH_3-CH_2-OH and	$HO-CH_2-CH_2-OH$
Principal group:	$-OH$	-ol
Parent hydride:	CH_3-CH_3	Ethane
Parent hydride + one principal group:	CH_3-CH_2-OH	Ethanol
Parent hydride + two principal groups:	$\underset{2}{H}O-CH_2-\underset{1}{C}H_2-OH$	Ethane-1,2-diol

[52] The position of hydro prefixes in the precedence order is under consideration by the Commission[49].

72

R-4.2.2

$$HO-CH_2-CH_2-CH_2-CH_2-CH_2-\overset{\overset{\displaystyle O}{\|}}{C}-CH_3$$

Principal group:	$> (C) = O$	-one
Parent hydride:	$CH_3-CH_2-CH_2-CH_2-CH_2-CH_2-CH_3$	Heptane
Parent hydride + principal group:	$\underset{7\ \ \ \ 6\ \ \ \ \ 5\ \ \ \ \ 4\ \ \ \ \ 3\ \ \ \ \ 2\ \ \ 1}{CH_3-CH_2-CH_2-CH_2-CH_2-\overset{\overset{\displaystyle O}{\|}}{C}-CH_3}$	Heptan-2-one
Substituent:	-OH	Hydroxy-
Compound name:	7-Hydroxyheptan-2-one	

R-4.2.3

$$\underset{}{CH_2=CH-CH_2-\overset{\overset{\displaystyle OH}{|}}{CH}-CH_3}$$

Principal group:	-OH	-ol
Parent hydride:	$CH_3-CH_2-CH_2-CH_2-CH_3$	Pentane
Parent hydride + principal group:	$\underset{5\ \ \ \ 4\ \ \ \ \ 3\ \ \ \ \ 2\ \ \ \ 1}{CH_3-CH_2-CH_2-\overset{\overset{\displaystyle OH}{|}}{CH}-CH_3}$	Pentan-2-ol
Subtractive modification:	$-2H$	-en-
Compound name:	Pent-4-en-2-ol	

R-4.2.4

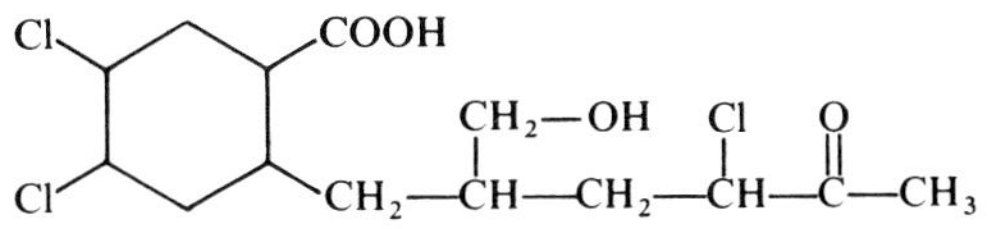

Principal group (from Table 5):	-COOH -(C)OOH	-carboxylic acid -oic acid
Parent hydride (R-2.3 and R-4.1b, c) (part of the structure bearing the principal group):	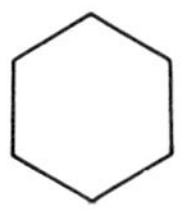	Cyclohexane
Parent hydride + principal group:	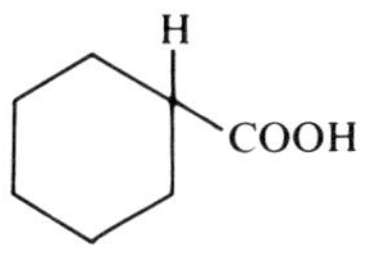	Cyclohexanecar-boxylic acid

Example continued on p. 74.

Substituents:	–Cl	Chloro-
Parent substituent:	–CH$_2$–CH$_2$–CH$_2$–CH$_2$–CH$_2$–CH$_3$ 1 2 3 4 5 6	Hexyl-
Substituents:	–Cl	Chloro-
	=O	Oxo-
	–CH$_2$–OH	
Secondary parent substituent:	–CH$_3$	Methyl-
Substituent:	–OH	Hydroxy-
Secondary substituent name:		Hydroxymethyl-
Substituent name:		4-Chloro-2-(hydroxymethyl)-5-oxohexyl-
Compound name:	4,5-Dichloro-2-[4-chloro-2-(hydroxymethyl)-5-oxohexyl]cyclo-hexane-1-carboxylic acid	

R-4.2.5

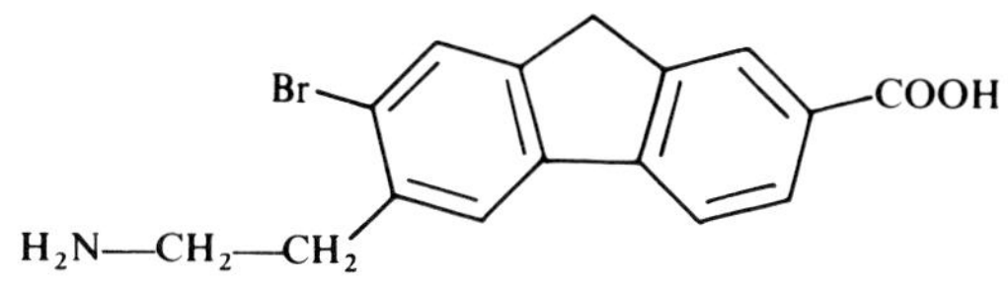

Principal group:	–COOH	-carboxylic acid
Parent hydride:		9H-Fluorene
Parent hydride + principal group:		9H-Fluorene-2-carboxylic acid
Substituents:	–Br	Bromo-
	–CH$_2$–CH$_2$–NH$_2$	
Secondary parent substituent:	–CH$_2$–CH$_3$ 1 2	Ethyl-
Substituent:	–NH$_2$	Amino-
Secondary substituent name:		2-Aminoethyl-

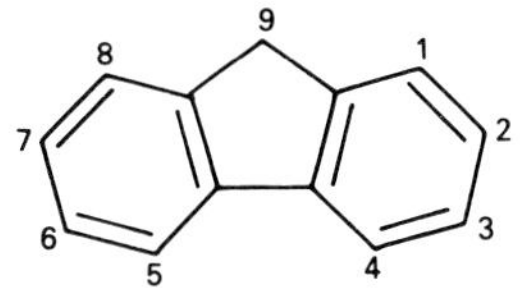

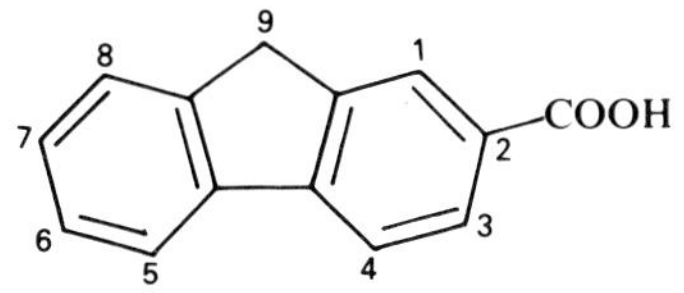

Compound name: 6-(2-Aminoethyl)-7-bromo-9H-fluorene-2-carboxylic acid

R-4.2.6

$$HOOC-[CH_2]_2-O-CH_2-CH_2-O-CH_2-CH_2-O-[CH_2]_2-COOH$$

In this example, the nomenclature of assemblies of identical units (see R-1.2.8) may be used. Two identical units, $-CH_2-COOH$, are linked by the symmetrical connecting substituent group $-O-CH_2-CH_2-O-CH_2-CH_2-O-$.

Principal group:	$-COOH$	-carboxylic acid
	$-(C)OOH$	-oic acid
Parent hydride:	$CH_3-CH_2-CH_3$	Propane
Parent hydride + principal group:	CH_3-CH_2-COOH	Propanoic acid
Multiplicative connecting group:	$-O-CH_2-CH_2-O-CH_2-CH_2-O-$	
Components:	$-O-$	-oxy-
	$-CH_2-CH_2-$	-ethylene-
Connecting substituent name:		Oxybis(ethyleneoxy)-

Compound name: 3,3'-[Oxybis(ethyleneoxy)]dipropanoic acid

Note: The numbering of the unit including the principal characteristic group (COOH) is retained.

R-4.2.7

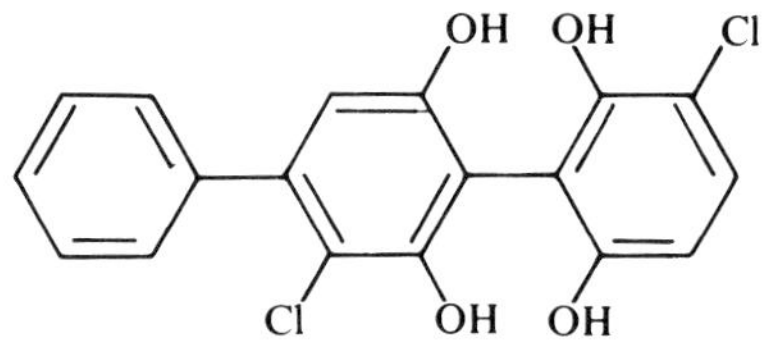

| Principal group: | $-OH$ | -ol |
| Parent hydride (see R-2.4.4): | | 1,1':4',1''-Terphenyl |

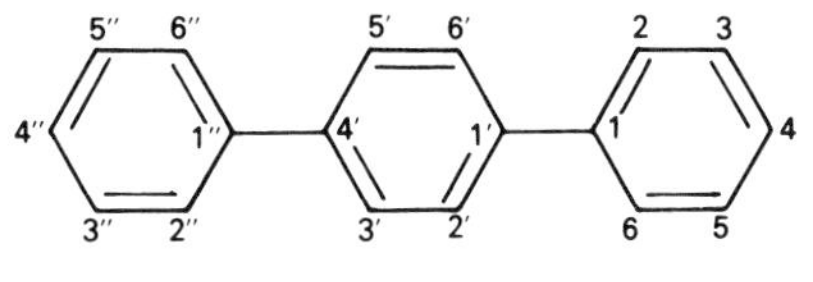

1,1':4',1''-Terphenyl-2,2',6,6'-tetrol

Parent hydride + principal group:

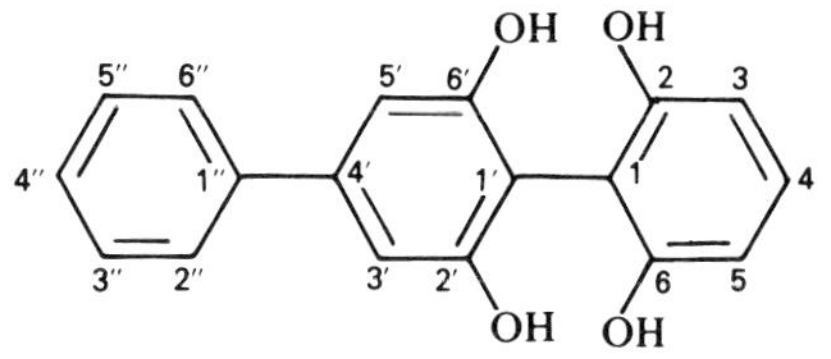

| Substituent: | $-Cl$ | Chloro- |
| Compound name: | 3,3'-Dichloro-1,1':4',1''-terphenyl-2,2',6,6'-tetrol | |

R-4.2.8

$$CH_3-CH_2-O-[CH_2]_2-O-[CH_2]_2-O-[CH_2]_2-O-[CH_2]_2-COOH$$

In this example, the replacement operation (see R-1.2.2) may be applied. For acyclic systems, the replacement operation is applied to the name of the parent hydride after consideration of the principal characteristic group(s). Accordingly, the principal characteristic group is preferred to the heteroatoms for low locants.

Principal group:	$-COOH$	-carboxylic acid
	$-(C)OOH$	-oic acid
Parent hydride:	$CH_3-[CH_2]_{13}-CH_3$	Pentadecane
Parent hydride + principal group:	$CH_3-[CH_2]_{13}-COOH$	Pentadecanoic acid
Replacement 'a' prefix:	$-O-$	oxa-
Compound name:	4,7,10,13-Tetraoxapentadecanoic acid	

R-4.2.9

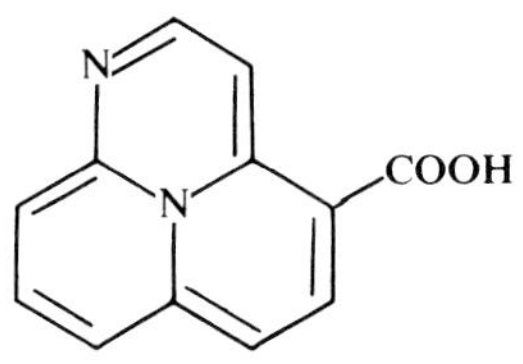

In this example, the replacement operation (see R-1.2.2) must be applied. However, for heterocyclic compounds, the replacement operation is applied to the parent name of the cyclic hydrocarbon before consideration of indicated hydrogen or the principal characteristic group. Accordingly, the heteroatoms are preferred to indicated hydrogen and to the principal characteristic group for low locants.

Principal group:	$-COOH$	-carboxylic acid
Parent hydrocarbon:	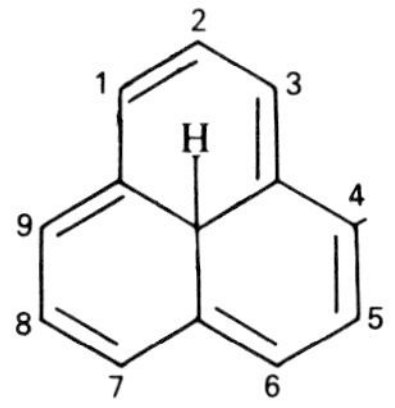	Phenalene (9b*H*-form shown)
Replacement 'a' prefix:	$-N<$	aza-
Parent heterocycle:	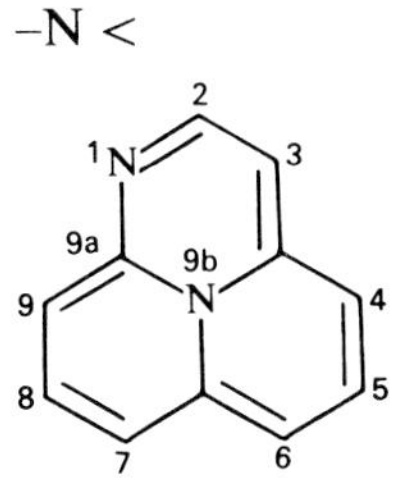	1,9b-Diazaphenalene
Compound name:	1,9b-Diazaphenalene-4-carboxylic acid	

R-5 Applications to Specific Classes of Compounds

R-5.0   INTRODUCTION

The following recommendations illustrate how the general principles set out in the preceding sections are applied to various types of compounds. More detailed recommendations will appear in a forthcoming publication (see also the 1979 edition of the IUPAC *Nomenclature of Organic Chemistry*[1]).

R-5.1   BINARY HYDRIDES AND RELATED PARENT HYDRIDES

Binary hydrides for which a parent hydride name is not available (see R-2) have a name that consists of a parent hydride name and appropriate substitutive prefixes.

R-5.1.1   **Hydrocarbons**

Substituted hydrocarbons are named as derivatives of carbon parent hydrides (see R-2.2, 2.3, and 2.4). Some trivial names are retained (see R-9.1, Tables 19(a), p. 162; 20, p. 164 and 21, p. 166).

Examples:

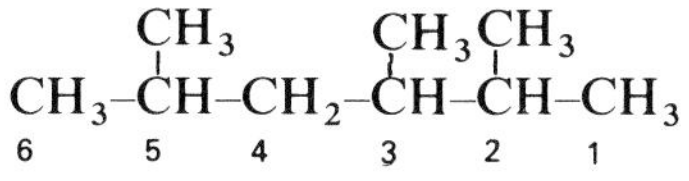

2,3,5-Trimethylhexane

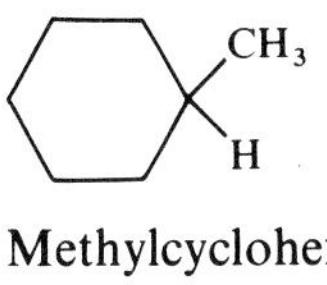

Methylcyclohexane

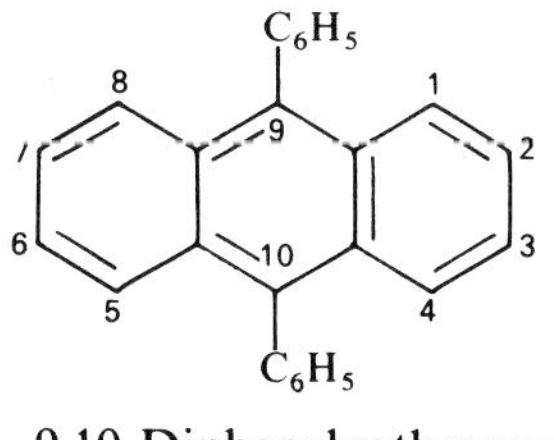

9,10-Diphenylanthracene

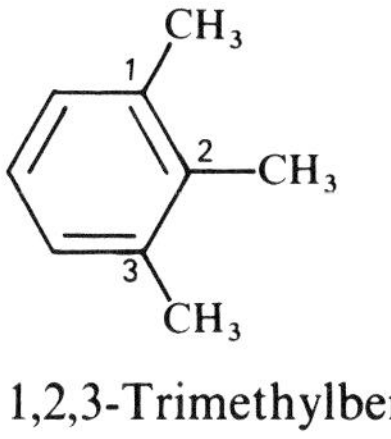

1,2,3-Trimethylbenzene

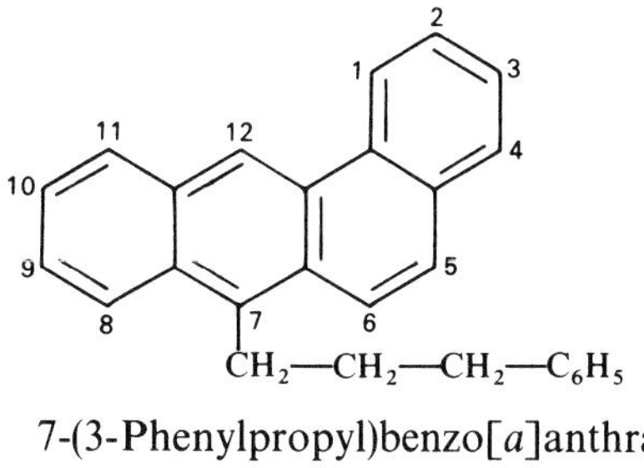

7-(3-Phenylpropyl)benzo[*a*]anthracene

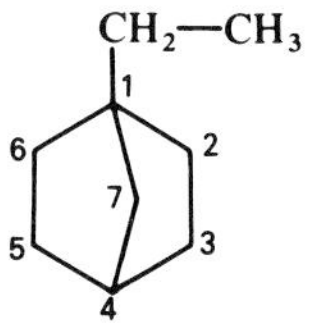

1-Ethylbicyclo[2.2.1]heptane

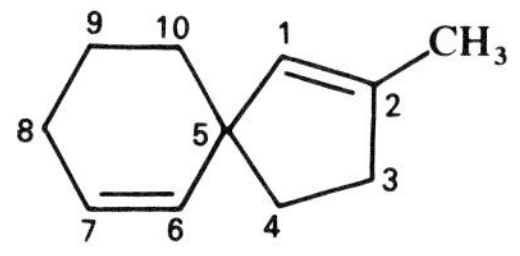

2-Methylspiro[4.5]deca-1,6-diene

R-5.1.2 **Chalcogen hydrides**

Derivatives of the parent hydrides HOH, HOOH, etc., and their sulfur, selenium, and tellurium analogues are treated in R-5.5.

R-5.1.3 **Hydrides of the group 15 elements**

R-5.1.3.1 Derivatives of the nitrogen hydrides may be named as amines, amides, imines, or hydrazides (see R-5.4 and R-5.7.8); or substitutively on the basis of parent hydrides (see R-2.2, R-2.3, and 2.4).

Examples:

$(C_6H_5)_2N-N=N-N=N-N(C_6H_5)_2$
$\quad\quad\quad\quad 6\quad 5\quad 4\quad 3\quad 2\quad 1$

1,1,6,6-Tetraphenylhexaaza-2,4-diene

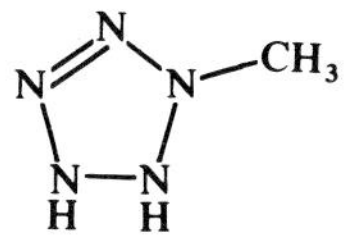

3-Methylcyclopentaaz-1-ene (R-2.3.2)
1-Methyl-2,3-dihyro-1*H*-pentazole
(R-2.3.3.3)

R-5.1.3.2 Organic derivatives of the trivalent phosphorus, arsenic, antimony, and bismuth parent hydrides are named substitutively on the basis of parent hydride names such as phosphane, diphosphane, arsane, stibane, bismuthane, etc. (see R-2.1 and R-2.2.2). In these recommendations, the names phosphine, arsine, stibine, and bismuthine are not encouraged.

Examples:

$CH_3-CH_2-PH_2$
Ethylphosphane

$(CH_3)_3Bi$
Trimethylbismuthane

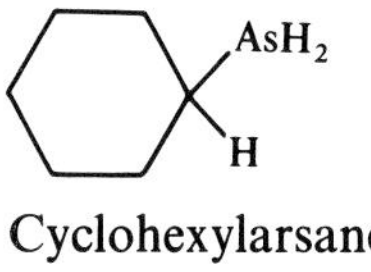

Cyclohexylarsane

$(C_6H_5)_2P-P(C_6H_5)_2$
Tetraphenyldiphosphane

$(CH_2=CH)_3Sb$
Trivinylstibane

R-5.1.3.3 Organic derivatives of pentavalent phosphorus, arsenic, antimony, and bismuth are named either on the basis of parent hydride names such as λ^5-phosphane, λ^5-arsane, and

78

$1\lambda^5, 2\lambda^5$-diphosphane, etc., or on the basis of parent hydride names such as phosphorane and arsorane, the former method being preferred.

Examples:

$(C_6H_5)_5P$
Pentaphenyl-λ^5-phosphane
Pentaphenylphosphorane

$(CH_3)_5As$
Pentamethyl-λ^5-arsane
Pentamethylarsorane

$$\begin{array}{c} CH_2-CH_3 \\ | \\ PH_4-PH_3-PH_2-PH_4 \\ \;\;4\quad\;\;3\quad\;\;2\quad\;\;1 \end{array}$$

2-Ethyl-$1\lambda^5,2\lambda^5,3\lambda^5,4\lambda^5$-tetraphosphane

R-5.1.4   Silicon parent hydrides

R-5.1.4.1   *Silanes.* Silicon analogues of hydrocarbons are called 'silanes' generically and are named analogously to the corresponding hydrocarbons.

Examples:

$$\begin{array}{c} SiH_3-SiH_2\;\;SiH_3 \\ \quad\;\;\;|\qquad\;\; | \\ SiH_3-SiH_2-SiH—SiH-SiH_3 \\ \;5\quad\;\;4\quad\;\;3\quad\;\;2\quad\;1 \end{array}$$

3-Disilanyl-2-silylpentasilane[53]

$$\begin{array}{c} CH_3 \\ | \\ SiH_3-SiH_2-SiH-SiH_3 \\ \;4\quad\;\;3\quad\;\;2\quad\;1 \end{array}$$

2-Methyltetrasilane

R-5.1.4.2   *Heterogeneous silicon hydrides: siloxanes and analogues.* Silicon compounds having the general formula $SiH_3-[O-SiH_2]_n-O-SiH_3$ are called 'siloxanes' generically and are named on the basis of parent hydride names such as 'disiloxane', 'trisiloxane', etc., according to the number of silicon atoms $(n + 2)$ in the chain. Sulfur, selenium, tellurium, and saturated nitrogen analogues are named in the same way as 'silathianes', 'silaselenanes', 'silatelluranes' and 'silazanes', respectively (see also R-2.2.3.2).

Examples:

$$\begin{array}{c} SiH_3-O-SiH_2-O-SiH_2-O-SiH_3 \\ \;7\quad\;\;6\;\;5\qquad4\;3\qquad2\;\;1 \end{array}$$

Tetrasiloxane

$$\begin{array}{c} SiH_3-NH-SiH_2-NH-Si(CH_3)_3 \\ \;5\qquad4\qquad3\qquad2\qquad1 \end{array}$$

1,1,1-Trimethyltrisilazane

$$\begin{array}{c} CH_3 \\ | \\ SiH_3-S-SiH-S-SiH_3 \\ \;5\qquad4\;\;3\quad\;2\;\;1 \end{array}$$

3-Methyltrisilathiane

---

[53] The abbreviated form 'silyl' is used instead of 'silanyl' in order to distinguish between two silyl (SiH_3-) groups, 'disilyl', and one 'disilanyl' (SiH_3-SiH_2-) group, without using the multiplicative prefix 'bis-'.

Monocyclic siloxanes having the general formula $\overparen{[O-SiH_2]_n}$ are named 'cyclo-trisiloxane', 'cyclotetrasiloxane', etc., according to the number of silicon atoms (n) present. Sulfur, selenium, tellurium, and saturated nitrogen analogues are named as 'cyclosilathianes', 'cyclosilaselenanes', etc., respectively (see also R-2.3.3.2).

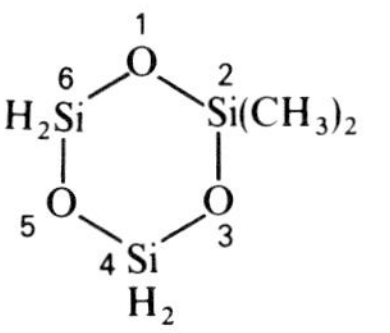

2,2-Dimethylcyclotrisiloxane

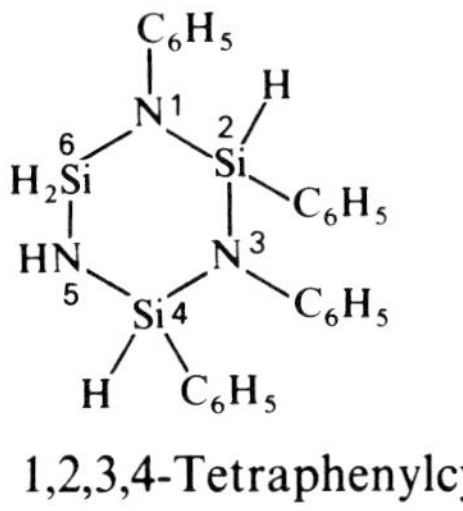

1,2,3,4-Tetraphenylcyclotrisilazane

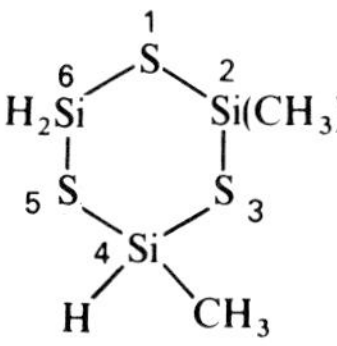

2,2,4-Trimethylcyclotrisilathiane

Bi- and polycyclic siloxanes, silathianes, silaselenanes, silatelluranes, and silazanes are named by citing a prefix defining the ring structure, such as 'bicyclo[3.3.1]' and 'spiro[5.7]', followed by a numerical term giving the number of silicon atoms present in the ring system, the whole being attached to 'siloxane', 'silathiane', 'silaselenane', 'silatellurane', 'silazane', as appropriate. In the case of bi- and polycyclic silazanes, the prefix *1Si-* or *1N-* is used when necessary to indicate the atom at the bridgehead that is numbered 1 (see also R-2.4.2). Additional details for naming these compounds are given in the 1979 edition of the IUPAC *Nomenclature of Organic Chemistry*[1] (see Rule D-6, pp. 411–416).

Examples:

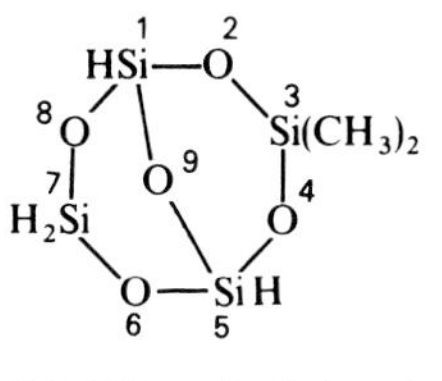

3,3-Dimethylbicyclo[3.3.1]tetrasiloxane

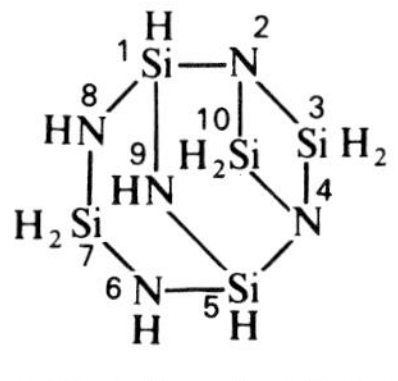

1Si-Tricyclo[3.3.1.1^{2,4}]pentasilazane

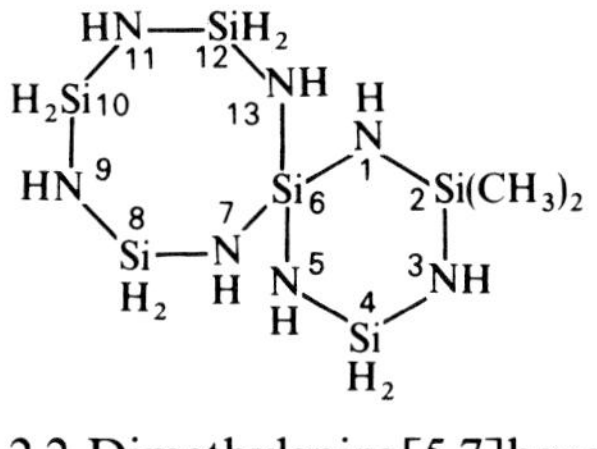

2,2-Dimethylspiro[5.7]hexasilazane

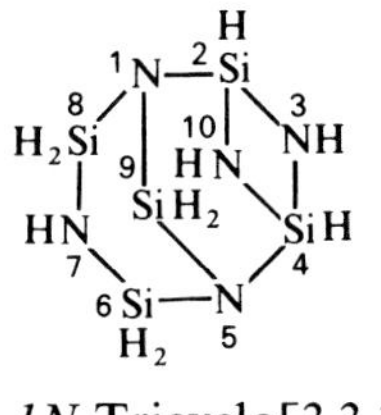

1N-Tricyclo[3.3.1.1^{2,4}]pentasilazane

R-5.2   ORGANOMETALLIC COMPOUNDS
             Although many organometallic compounds exist in associated molecular forms and
             contain structural solvent, their names are generally based on the stoichiometric
             compositions of the compounds, the solvent, if any, being ignored[54].

             Examples:

$(C_4H_9Li)_n$ $[(C_2H_5)_3Al]_2$
Butyllithium (see R-5.2.2)   Triethylaluminium (see R-5.2.2)

R-5.2.1   **Organometallic compounds of antimony, bismuth, germanium, tin, and lead**
             These may be named substitutively on the basis of the appropriate parent hydride (see
             R-2.1)[55]. In these recommendations, this method is preferred.

             Examples:

$(C_2H_5)_3Bi$ $(C_6H_5)_2(C_2H_5)(CH_3)Ge$
Triethylbismuthane   Ethyl(methyl)diphenylgermane

$(C_6H_5)_4Pb$
Tetraphenylplumbane $(CH_3)_5Sb$
                                            Pentamethyl-λ^5-stibane
$(C_6H_5)_2SnH_2$ Pentamethylstiborane
Diphenylstannane

R-5.2.2   **Organometallic compounds in which the metal is bound only to carbon atoms of**
             organic groups and hydrogen
             These may be named by citing prefixes denoting the organic groups and the prefix
             'hydrido-' denoting hydrogen atoms, if any, in alphabetical order, followed by the name
             of the metal. The presence of any hydrogen atoms attached to the metal atom must
             always be so indicated. There is no space between the group names and that of the metal.

             Note: For antimony, bismuth, germanium, tin, and lead, substitutive names based on the
             corresponding parent hydride, where appropriate, are preferred in these recommenda-
             tions (see R-5.2.1).

             Examples:

$(CH_3Li)_n$
Methyllithium

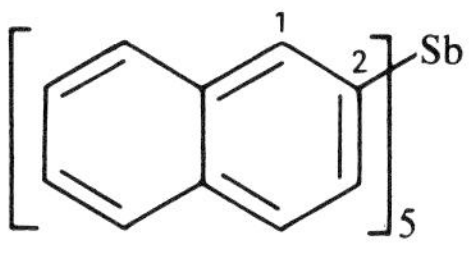

                                            Penta-2-naphthylantimony
                                            Penta-2-naphthyl-λ^5-stibane (preferred)
                                            Penta-2-naphthylstiborane (R-5.2.1)

$CH_2=CHNa$ C_2H_5BeH
Vinylsodium   Ethylhydridoberyllium

$(C_4H_9)_3GeH$ $(C_6H_5)(C_6H_5CH_2)AuH$
Tributylhydridogermanium   Benzylhydridophenylgold
Tributylgermane (preferred) (R-5.2.1)

[54] See IUPAC *Nomenclature of Inorganic Chemistry*[3], Recommendation I.10.9, pp. 198–205.
[55] See IUPAC *Nomenclature of Inorganic Chemistry*[3], Recommendation I.7.2, pp. 82–96.

R-5.2.3 **Organometallic compounds with anionic ligands**
These may be named by citing the name(s) of the organic group(s), in alphabetical order, followed by the name of the metal and the anion name(s) in alphabetical order if more than one. The names of the anions are separated from the name of the metal, and from each other, by spaces. Hydrogen may also be included in alphabetical order with the organic groups. They may also be named by citing the name(s) of all the attached group(s), including hydrogen (ligand name 'hydrido') and anionic groups expressed as ligands, in alphabetical order, followed by the name of the metal (see *Nomenclature of Inorganic Chemistry*[3], Recommendation I-10).

Note: For antimony, bismuth, germanium, tin, and lead, substitutive names based on the corresponding parent hydride, where appropriate, are preferred in these recommendations (see R-5.2.1).

Examples:

$(CH_3MgI)_n$
Methylmagnesium iodide
Iodo(methyl)magnesium

$(C_6H_5)_2SbCl$
Chlorodiphenylstibane (R-5.2.1)
Diphenylantimony chloride
Chlorodiphenylantimony

CH_3SnH_2Cl
Chloro(methyl)stannane (R-5.2.1)
Methyltin chloride dihydride
Dihydridomethyltin chloride
Chlorodihydridomethyltin

R-5.3 HALOGEN, NITRO, NITROSO, AZO, DIAZO AND AZIDO COMPOUNDS

R-5.3.1 **Halogen compounds**
Halogen compounds may be named systematically according to two systems. Some trivial names are retained (see R-9.1, Table 32, p. 180). Substitutive names are formed by adding the prefixes 'fluoro-', 'chloro-', 'bromo-', or 'iodo-' to the name of the parent compound.

Examples:

$$\underset{6\quad\ 5\quad\ 4\quad\ 3\quad\ 2\quad\ 1}{CH_3-CH_2-CH_2-CH_2-\overset{\overset{\displaystyle Cl}{|}}{CH}-CH_3}$$

2-Chlorohexane

$$\underset{2\qquad\ 1}{Cl-CH_2-CH_2-Br}$$

1-Bromo-2-chloroethane

3-Bromopyridine

$$\underset{5\qquad\qquad\qquad\ 1}{(CH_3)_3Si-[SiH_2]_3-SiCl_3}$$

1,1,1-Trichloro-5,5,5-trimethylpentasilane

$NF_2-CO-NF_2$
Tetrafluorourea

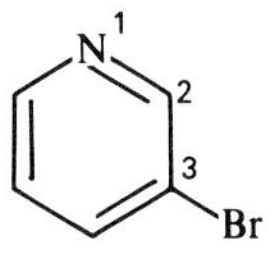

Functional class names are formed by naming the organic 'groups' followed by the class name 'fluoride', 'chloride', 'bromide', or 'iodide', preceded, if necessary, by a numerical prefix, as a separate word.

Examples:

CH_3–I
Methyl iodide

$(CH_3)_3C$–Cl
tert-Butyl chloride

C_6H_5–CH_2–Br
Benzyl bromide

Br–CH_2–CH_2–Br
Ethylene dibromide

R-5.3.2   Nitro and nitroso compounds

Compounds containing a –NO_2 group or a –NO group are named only by means of the prefixes 'nitro-' or 'nitroso-', respectively.

Examples:

CH_3–NO_2
Nitromethane

1,2-Dinitrobenzene (preferred)
o-Dinitrobenzene

1-Nitronaphthalene

C_6H_5–NO
Nitrosobenzene

Compounds containing the group =N(O)OH may be named as derivatives of the functional parent compound azinic acid $H_2N(O)OH$ or by using the prefix hydroxy-nitroryl (see Table 7)[56].

Examples:

CH_2=N(O)–OH
Methylideneazinic acid

2-(Hydroxynitroryl)cyclohexane-1-carboxylic acid

R-5.3.3   Azo, azoxy, diazo, and related compounds

R-5.3.3.0   *Diazenes.* The parent structure HN=NH is named 'diazene'[57,58] and the groups derived from it, i.e., HN=N– and –N=N–, are named systematically as diazenyl and diazenediyl, respectively[59] (see R-3.1.4).

[56] The prefix '*aci*-nitro-' is not included in these recommendations.

[57] International Union of Pure and Applied Chemistry. Inorganic Chemistry Division. Commission on Nomenclature of Inorganic Chemistry, 'The Nomenclature of Hydrides of Nitrogen and Derived Cations, Anions, and Ligands', *Pure Appl. Chem.*, **54**, 2545–2552 (1982).

[58] This compound has also been known as diimide.

[59] In previous editions of the IUPAC *Nomenclature of Organic Chemistry*[1], the prefix name for the –N=N– group was 'azo-'.

R-5.3.3.1   *Azo compounds.* Compounds with the general structure R–N=N–R', where R and R' may be alike or different, are known generically as 'azo compounds'. However, these compounds may be more systematically named substitutively as derivatives of the parent hydride diazene[60].

Examples:

CH$_3$–N=N–CH$_3$
Dimethyldiazene
(traditionally Azomethane)

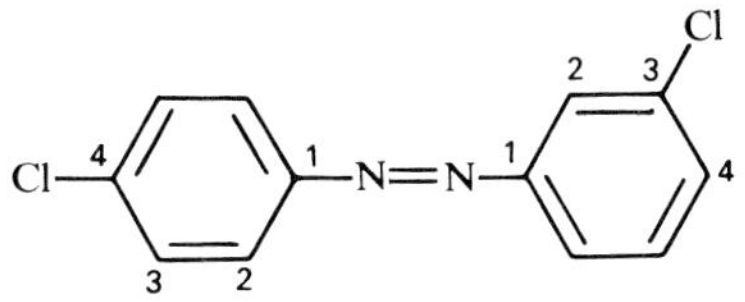

(3-Chlorophenyl)(4-chlorophenyl)diazene
(traditionally 3,4'-Dichloroazobenzene)

C$_6$H$_5$–N=N–C$_6$H$_5$
Diphenyldiazene
(traditionally Azobenzene)

CH$_2$=CH–N=N–CH$_3$
Methyl(vinyl)diazene
(traditionally Etheneazomethane)

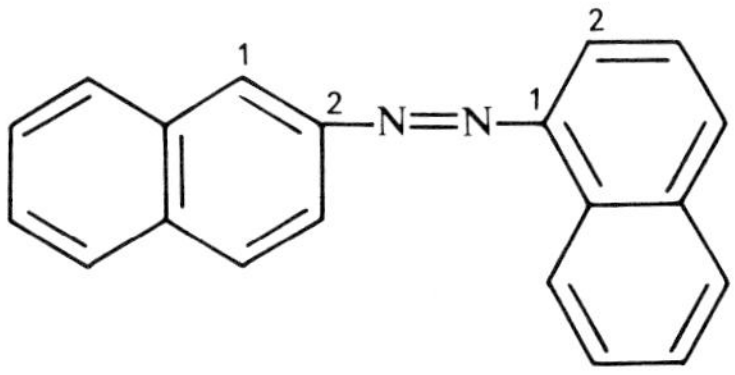

(1-Naphthyl)(2-naphthyl)diazene
(traditionally 1,2'-Azonaphthalene)

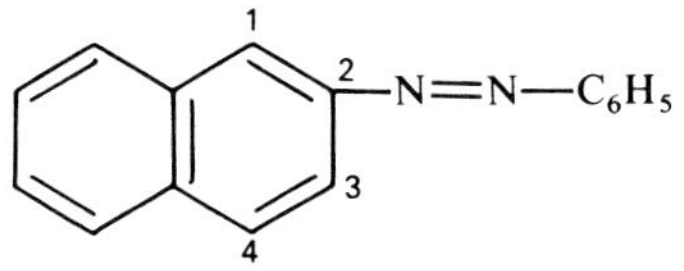

(2-Naphthyl)phenyldiazene
(traditionally Naphthalene-2-azobenzene)

A monoazo compound with the general structure R–N=N–R' in which R is substituted by a principal characteristic group is named on the basis of the parent hydride, RH, substituted by an organyldiazenyl group, R'–N=N–. If both R and R' are substituted by the same number of the principal characteristic group, a multiplicative name (see R-0.2.3.3.10) may be used.

Examples:

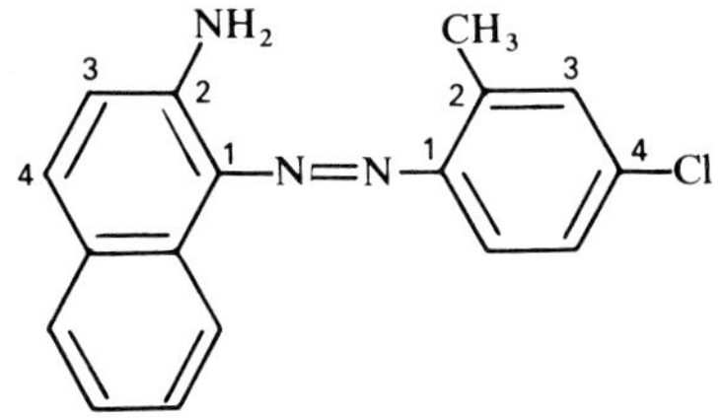

1-[(4-chloro-2-methylphenyl)diazenyl]naphthalen-2-amine

[60] According to the traditional method for naming azo compounds, as described in previous editions of the IUPAC *Nomenclature of Organic Chemistry*[1], symmetrical monoazo compounds, R–N=N–R, are named by adding the prefix 'azo-' to the name of the parent hydride, RH, and unsymmetrical monoazo compounds, R–N=N–R', are named by placing '-azo-' between the names of the parent hydrides RH and R'H. Unsymmetrical compounds in which the parent hydride, RH, bears a principal characteristic group are named as derivatives of that parent hydride substituted by an organylazo group, R'–N=N–. Details of the application of these traditional rules, which are recognized as an acceptable alternative, are given in the 1979 edition of the IUPAC *Nomenclature of Organic Chemistry*[1] (see Rule C-9.1, pp. 277–283).

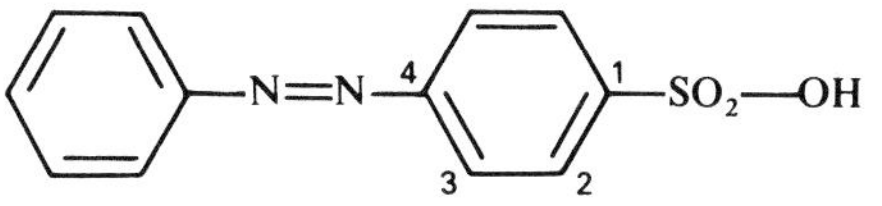

4-(Phenyldiazenyl)benzenesulfonic acid
[traditionally 4-(Phenylazo)benzenesulfonic acid]

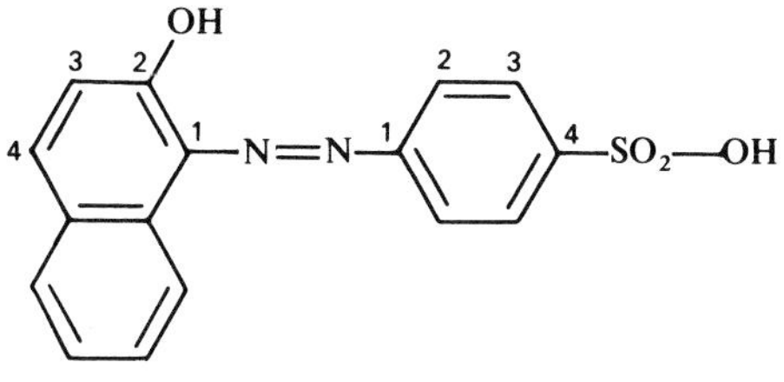

4-[(2-Hydroxy-1-naphthyl)diazenyl]benzenesulfonic acid
[traditionally 4-[(2-Hydroxy-1-naphthyl)azo]benzenesulfonic acid]

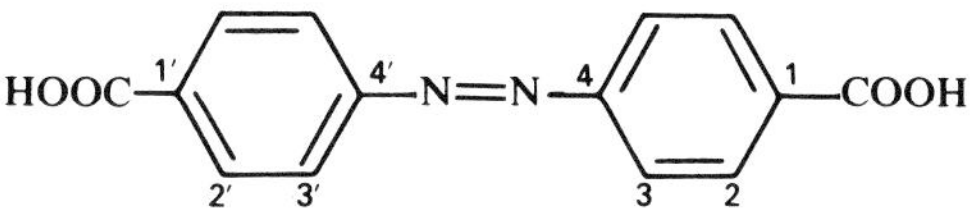

4,4'-Diazenediyldibenzoic acid (traditionally 4,4'-Azodibenzoic acid)

Bisazo compounds and more complex analogues are named on the basis of the parent structure "diazene", in the absence of a more preferred parent compound.

Example:

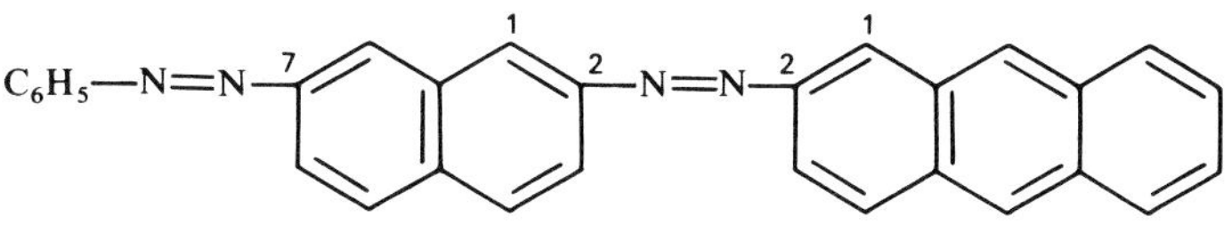

(2-Anthryl)[(7-phenyldiazenyl)-2-naphthyl)]diazene
(traditionally Anthracene-2-azo-2'-naphthalene-7'-azobenzene)

R-5.3.3.2   *Azoxy compounds.* N-Oxides of azo compounds having the general structure $R-N_2(O)-R$ or $R-N_2(O)-R'$ are known generically as 'azoxy compounds' and are named by adding the separate word 'oxide' to the name of the corresponding azo compound (see R-5.3.3.1)[61]. In an unsymmetrical compound, the position of the azoxy oxygen atom is expressed by the locant 1 or 2[62].

Examples:

$C_6H_5-N_2(O)-C_6H_5$
Diphenyldiazene oxide
(traditionally Azoxybenzene)

---

[61] In the previous edition of the IUPAC *Nomenclature of Organic Chemistry*[1], azoxy compounds were named in the same way as azo compounds using 'azoxy' in place of 'azo'.
[62] A notation -NNO-, -ONN-, or -NON- was used to express the position of the oxygen atom in the previous edition of the IUPAC *Nomenclature of Organic Chemistry*[1] (see Rule C-913.2, pp. 283–284).

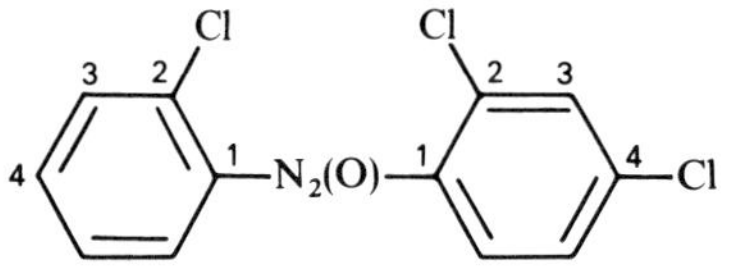

(2-Chlorophenyl)(2,4-dichlorophenyl)diazene oxide
(the position of the oxide is unknown)
(traditionally 2,2',4-Trichloroazoxybenzene)

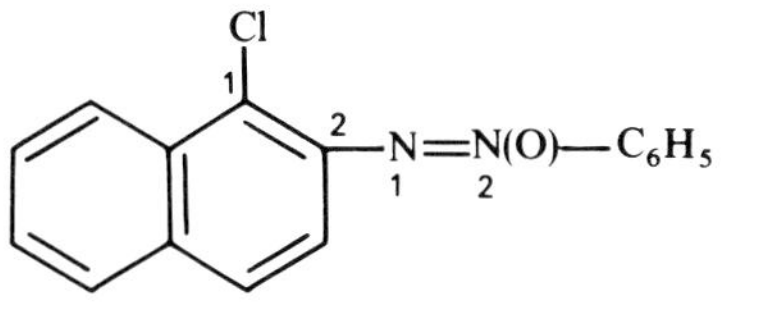

1-(1-Chloro-2-naphthyl)-2-phenyldiazene 2-oxide
(traditionally 1-Chloronaphthalene-*NNO*-azoxybenzene

An azoxy compound in which the general structure is R–N=N(O)–R' or R–N(O)=N–R', in which R is substituted by a principal characteristic group, is named on the basis of the parent hydride, RH, substituted by the R'-azoxy group in which the position of the oxygen atom is denoted by the prefix *NNO*-, *ONN*-, or *NON*-, as appropriate.

Example:

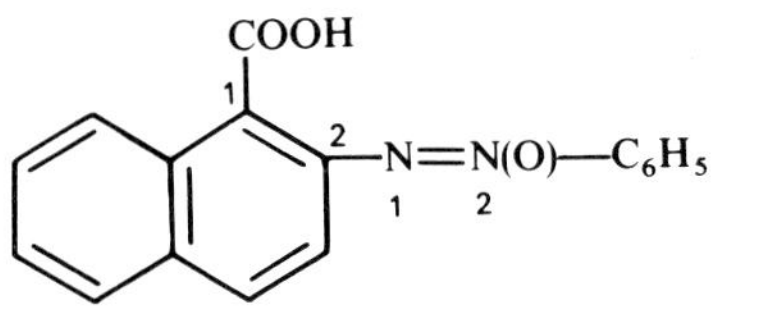

2-(Phenyl-*ONN*-azoxy)-1-naphthoic acid

Alternatively, the principal characteristic group may be ignored and the compound named substitutively on the basis of the parent hydride diazene. Accordingly, the above compound would be named 1-(1-carboxy-2-naphthyl)-2-phenyldiazene 2-oxide.

R-5.3.3.3   **Diazonium compounds.** Compounds with the general structure $R-N_2^+X^-$ are named by adding the suffix '-diazonium' to the name of the parent hydride RH followed by the name of the ion X^- as a separate word.

Examples:

$CH_3-CH_2-N_2^+Cl^-$
Ethanediazonium chloride

$C_6H_5-N_2^+Cl^-$
Benzenediazonium chloride

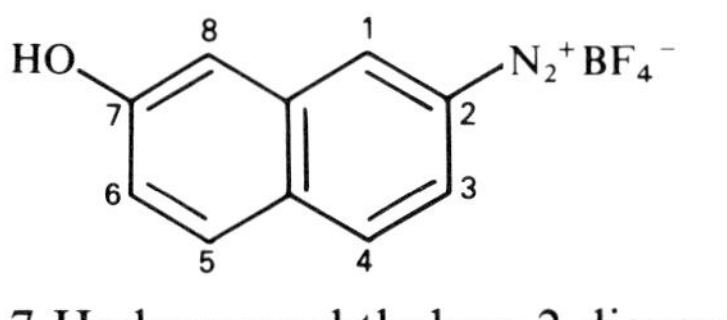

7-Hydroxynaphthalene-2-diazonium
tetrafluoroborate

R-5.3.3.4   **Azo compounds with the general structure R–N=N–X** should be named as derivatives of the parent structure diazene, HN=NH.

86

Examples:

$C_6H_5-N=N-OH$
Phenyldiazenol
(traditionally Benzenediazohydroxide)

$C_6H_5-N=N-O^- Na^+$
Sodium phenyldiazenolate
(traditionally Sodium benzenediazoate)

$C_6H_5-N=N-SO_3^- Na^+$
Sodium phenyldiazenesulfonate
(traditionally Sodium benzenediazosulfonate)

R-5.3.3.5   **Diazo compounds.** Compounds containing a group N_2 attached by one nitrogen atom to one carbon atom are named by adding a prefix 'diazo-' to the name of the parent hydride.

Examples:
CH_2N_2
Diazomethane

$N_2CH-CO-O-C_2H_5$
Ethyl diazoacetate

R-5.3.4   **Azides**
Compounds containing a group N_3 attached through one nitrogen atom to a parent hydride structure are named (a) in substitutive nomenclature by adding the prefix 'azido-' to the name of the parent hydride RH, or (b) in functional class nomenclature by citing the class name 'azide' as a separate word following the name of the group R.

Examples:
$C_6H_5-N_3$
(a) Azidobenzene
(b) Phenyl azide

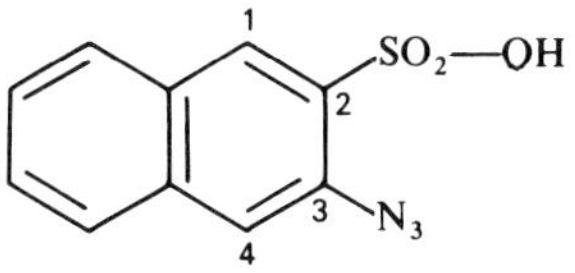

(a) 3-Azidonaphthalene-2-sulfonic acid

R-5.3.5   **Isodiazenes**
Compounds with the general structure $R_2N-N: \leftrightarrow R_2N^+=N^-$ may be named substitutively as derivatives of the parent 'radical' hydrazinylidene (see R-5.8.1.2) or on the basis of the trivial name isodiazene.

Example:
$(CH_3)_2N-N: \leftrightarrow (CH_3)_2N^+=N^-$

Dimethylisodiazene
Dimethylhydrazinylidene (R-5.8.1.2)

R-5.4   AMINES AND IMINES[63]
The generic name 'amine' is applied to compounds NH_2R, $NHRR'$, and $NRR'R''$, which are classified as primary, secondary, and tertiary amines, respectively.

[63] Salts of amines and imines are named by citing the name of the cation (see R-5.8.2) followed by the name of the anion as a separate word. Except for indexing and very complex cases, names such as .... amine hydrochloride are not encouraged.

R-5.4.1   **Primary amines**
            Primary amines, NH_2R, may be named according to one of three methods as follows:
            (a) by citing the name of the substituent group R as a prefix to the name of the parent
            hydride azane;
            (b) by adding the suffix '-amine' to the name of the parent hydride RH;
            (c) by adding '-amine' to the substituent name for the group R[64].

Examples:

$CH_3-CH_2-NH_2$
(a) Ethylazane
(b) Ethanamine
(c) Ethylamine

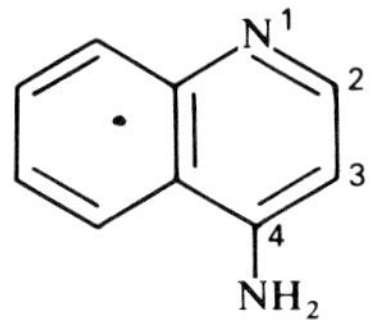

(a) 4-Quinolylazane
(b) Quinolin-4-amine
(c) 4-Quinolylamine

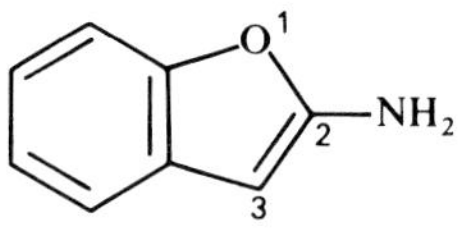

(a) 1-Benzofuran-2-ylazane
(b) 1-Benzofuran-2-amine
(c) 1-Benzofuran-2-ylamine

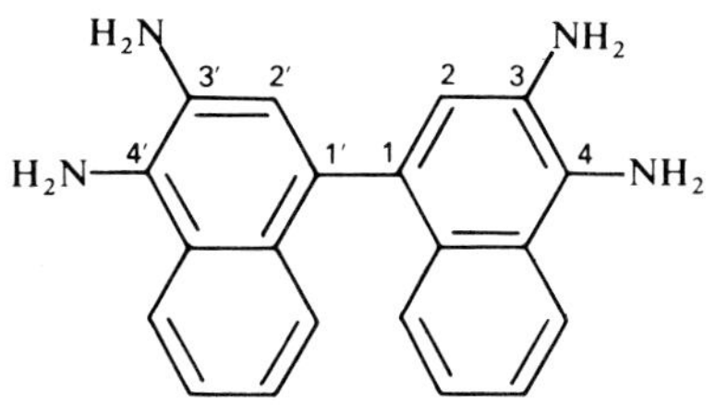

(a) [1,1'-Binaphthalene-3,3',4,4'-tetrayl]tetrakis(azane)
(b) [1,1'-Binaphthalene]-3,3',4,4'-tetramine
    [see R-0.1.7.1 (c)]
(c) [1,1'-Binaphthalene-3,3',4,4'-tetrayl]tetraamine

When it is not the principal characteristic group, the $-NH_2$ group is named by the prefix
'amino-'.

Example:

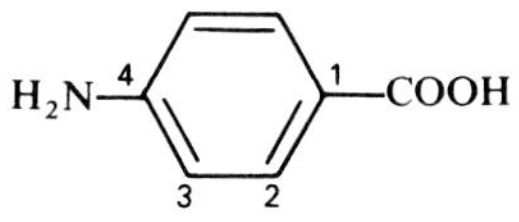

4-Aminobenzoic acid (preferred)
p-Aminobenzoic acid

Some trivial names are retained (see R-9.1, Table 24, p. 171).

R-5.4.2   **Secondary and tertiary amines**
            Symmetrical secondary and tertiary amines NHR_2 and NR_3 may be named according to
            two methods as follows:

---

[64] This has led to two treatments, one like (a) in which 'amine' is considered to be a parent hydride synonymous
with the parent hydride 'azane', and the second in which the name for $R-NH_2$, for example, ethylamine,
becomes a functional parent compound; the latter method is better illustrated with substituted or unsymmetrical secondary amines (see R-5.4.2). The first method is used by *Beilstein* and the second was used in Chemical
Abstracts index nomenclature prior to Volume **76** (1972). Both methods are illustrated in the 1979 edition of the
IUPAC *Nomenclature of Organic Chemistry*[1].

(a) by citing the name of the substituent group R, preceded by the numerical prefix 'di-' or 'tri-', respectively, as a prefix to the name of the parent hydride azane;

(b) by citing the name of the substituent group R, preceded by 'di-' or 'tri-', as appropriate, and followed directly, without a space, by the name 'amine'[65].

Examples:

$(C_6H_5)_2NH$
(a) Diphenylazane
(b) Diphenylamine

$(C_2H_5)_3N$
(a) Triethylazane
(b) Triethylamine

$(ClCH_2-CH_2)_2NH$
(a) Bis(2-chloroethyl)azane
(b) Bis(2-chloroethyl)amine
    2,2'-Dichlorodiethylamine

Unsymmetrical secondary and tertiary amines, $NHRR'$, NR_2R', and $NRR'R''$, may be named according to three methods as follows:

(a) as substituted derivatives of the parent hydride azane;

(b) as N-substituted derivatives of a primary amine RNH_2 or a secondary amine R_2NH;

(c) by citing the names of all substituent groups, R, R', or R'', preceded by appropriate numerical prefixes, and followed directly, without a space, by the class name 'amine'.

Substituent groups in names of unsymmetrical secondary and tertiary amines are ordered alphabetically.

Examples:

$ClCH_2-CH_2-NH-CH_2-CH_2-CH_3$
(a) (2-Chloroethyl)(propyl)azane
(b) *N*-(2-Chloroethyl)propan-1-amine
    N-(2-Chloroethyl)propylamine
(c) (2-Chloroethyl)(propyl)amine

$$CH_3-CH_2-CH_2-CH_2-\overset{\overset{\displaystyle CH_3}{|}}{N}-CH_2-CH_3$$
(a) Butyl(ethyl)methylazane
(b) *N*-Ethyl-*N*-methylbutan-1-amine
    N-Ethyl-*N*-methylbutylamine
(c) Butyl(ethyl)methylamine

R-5.4.3   **Imines**

Compounds having the general structure $R-CH=NR'$ or $RR''C=NR'$ (where R' may be equal to H) have been called generically 'aldimines' and 'ketimines', respectively. Imines with the general structure $R-CH=NH$ or $RR'C=NH$ may be named substitutively as '-ylidene' derivatives of the parent hydride azane or by replacing the final 'e', if present, of the name of the parent hydride $R-CH_3$ or $R-CH_2-R'$ with the suffix '-imine'. Compounds with the general structure $R-CH=N-R'$ or $RR''C=NR'$ may also be named as *N*-

---

[65] As for method (c) in R-5.4.1, this method has been applied in two ways, leading to names in which 'amine' is synonymous with 'azane', and to names that use the name for R_2NH as a functional parent (see second (b) names, in examples of this subsection).

substituted imines or as 'ylidene' derivatives of an amine R′–NH₂ (see R-5.4.1 and R-5.4.2)[66].

Note: The class terms 'aldimines' and 'ketimines' mentioned above have been derived from names such as 'benzaldehyde imine' or 'ethyl methyl ketone imine' in which 'imine' is a functional modifier.

Examples:

$CH_3–CH_2–CH_2–CH_2–CH_2–CH=NH$
Hexylideneazane
Hexan-1-imine
Hexylideneamine

$CH_3–CH=N–CH_3$
Ethylidene(methyl)azane
N–Methylethanimine
N-Methylethylideneamine

R-5.4.4   **Hydroxylamines**
Compounds with the general structure R–NH–OR′ are named substitutively on the basis of the functional parent compound 'hydroxylamine', using the locants *N*– and *O*– to distinguish between substitution on nitrogen or oxygen, or by attaching prefixes such as 'hydroxyamino-', 'alkoxyamino-', or '(aryloxy)amino-' to the name of the parent hydride R–H.

Examples:

$C_6H_5–NH–OH$
N-Phenylhydroxylamine

$CH_3–CO–O–NH–CH_3$
O-Acetyl-*N*-methylhydroxylamine

4-(Hydroxyamino)phenol

R-5.4.5   **Amine oxides**
Compounds with the general structure R₃NO, where the R groups may be alike or different, are named by adding the class name 'oxide' as a separate word after the name of the amine R₃N. Cyclic analogues are named in the same way, the position of the oxygen atom being indicated, where needed, by the locant of the ring atom (arabic numbers are preferred over capital italic element symbols as locants). Where needed, the R₂N(O)– group may be designated by a prefix derived from 'azinoyl-' (see Table 7, p. 65).

Examples:

$(CH_3)_3NO$
Trimethylazane oxide
Trimethylamine oxide

$(CH_3)_2N(O)–CH_2–C≡N$
(Dimethylazinoyl)acetonitrile

Pyridine 1-oxide (preferred)
Pyridine *N*-oxide

---

[66] In previous editions of the IUPAC *Nomenclature of Organic Chemistry*[1], imines were also named substitutively, as '-ylidene' derivatives of 'amine' (see footnote to method (c) for naming primary amines, R-5.4.1).

R-5.5 HYDROXY COMPOUNDS, THEIR DERIVATIVES
AND ANALOGUES

R-5.5.1 **Hydroxy compounds and analogues**

R-5.5.1.1 *Alcohols and phenols.* In substitutive nomenclature, the hydroxy group, $-OH$, as the principal characteristic group is indicated by adding a suffix, such as '-ol', '-diol', etc., as appropriate, to the name of the parent hydride with elision of a terminal 'e' before a following vowel.

Examples:

CH_3-OH
Methanol

$CH_3-\underset{4}{\overset{\overset{\displaystyle OH}{|}}{CH}}-\underset{3}{CH_2}-\underset{2}{CH_2}-\underset{1}{OH}$

Butane-1,3-diol

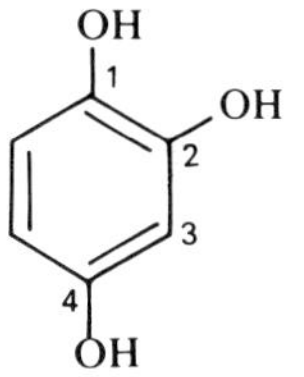

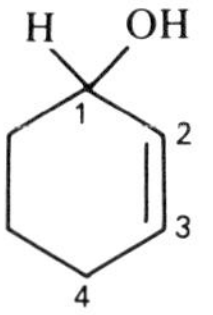

Cyclohex-2-en-1-ol

Bicyclo[4.2.0]octan-3-ol

Chrysen-1-ol

Benzene-1,2,4-triol

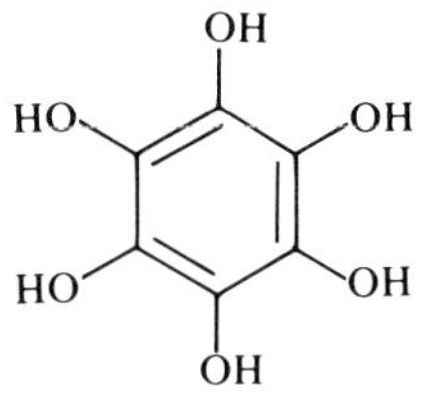

Benzenehexol [see R-0-1.7.1(c)]

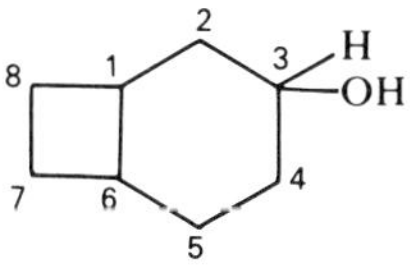

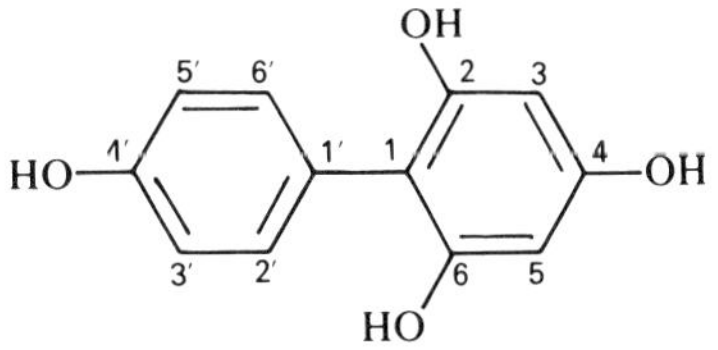

Biphenyl-2,4,4',6-tetrol [see R-0-1.7.1(c)]

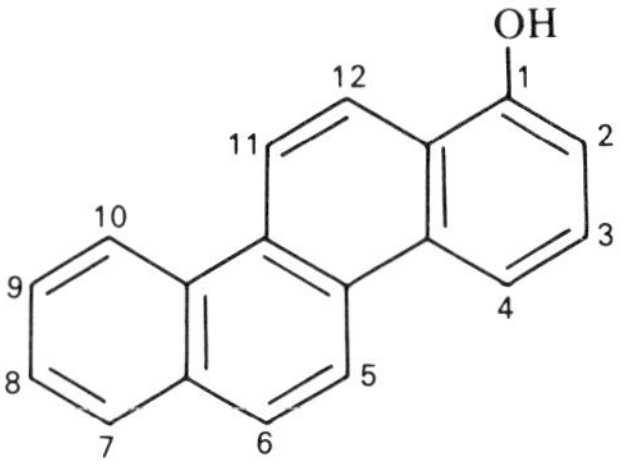

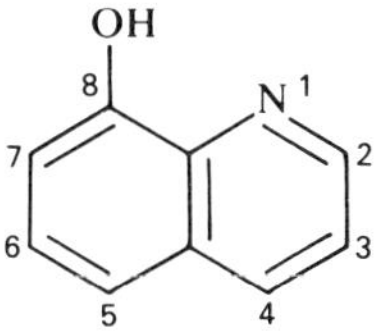

Quinolin-8-ol[67]

---

[67] The trivial name 'oxine', used as a ligand name in inorganic nomenclature of coordination compounds, is *not* recommended as it is the expected Hantzsch–Widman name for 2*H*-pyran.

When a group having priority for citation as the principal characteristic group is also present, hydroxy groups are indicated by the prefix 'hydroxy-'.

Examples:

OH
|
$CH_3-CH-CH_2-CH_2-CH_2-CH_2-CHO$
 7   6   5   4   3   2   1

6-Hydroxyheptanal

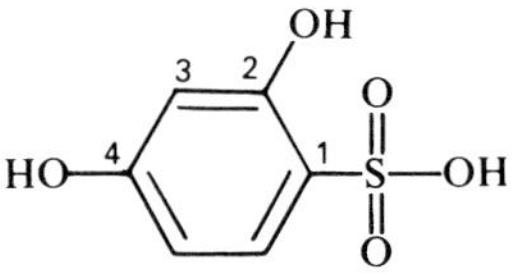

2,4-Dihydroxybenzenesulfonic acid

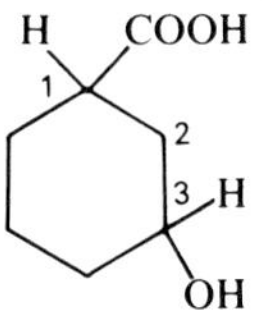

3-Hydroxycyclohexanecarboxylic acid

Functional class names for alcohols consist of the substituent prefix derived from the name of the corresponding parent hydride followed by the class name 'alcohol' cited as a separate word.

Examples:

CH_3-CH_2-OH
Ethyl alcohol

OH
|
$CH_3-CH_2-CH-CH_3$
sec-Butyl alcohol

The following contracted names are retained for the structures shown and their positional isomers.

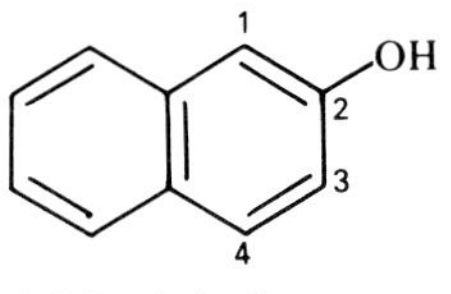

2-Naphthol

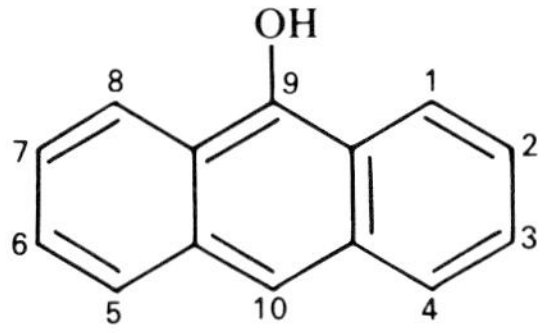

9-Anthrol

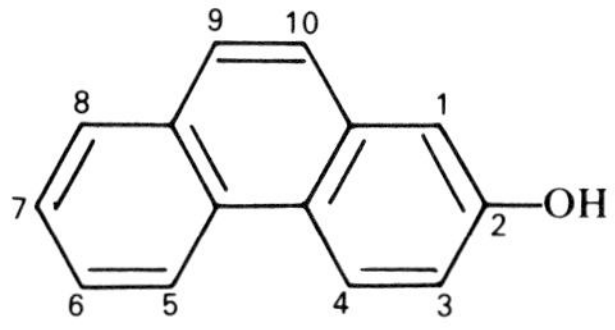

2-Phenanthrol

Some trivial names are retained (see R-9.1, Table 26(a), p. 173).

R-5.5.1.2   ***Sulfur, selenium, and tellurium analogues of alcohols and phenols*** are named in the same way by using suffixes such as '-thiol', '-selenol', and '-tellurol'[68], and prefixes such as

---

[68] The use of the prefixes 'thio-', 'seleno-', and 'telluro-' with a trivial name of a phenol to indicate the replacement of the hydroxy oxygen atom by sulfur, selenium, or tellurium, respectively, is not included in these recommendations.

'sulfanyl-', 'selanyl-", and 'tellanyl-' to designate the characteristic groups –SH, –SeH, and –TeH, respectively[69].

Examples:

CH$_3$–CH$_2$–SH
Ethanethiol

CH$_3$–CH$_2$–SeH
Ethaneselenol

C$_6$H$_5$–SH
Benzenethiol (**not** Thiophenol)

HS–CH$_2$–CH$_2$–COOH
 3 2 1

3-Sulfanylpropanoic acid

Naphthalene-2-thiol

4-Selanylbenzoic acid

R-5.5.2   **Substituent prefixes derived from alcohols, phenols, and their analogues**
Substituent prefix names for RO– groups are formed by adding 'oxy' to the substituent prefix name for the group R. Some contracted names, such as 'methoxy' and 'phenoxy', are retained (see R-9.1, Table 26(b), p. 173).

Examples:

CH$_3$–CH$_2$–CH$_2$–CH$_2$–CH$_2$–O–
Pentyloxy
Pentan-1-yloxy

2-Pyridyloxy
Pyridin-2-yloxy

Substituent prefix names for RS–, RSe–, and RTe– groups are formed by attaching the prefix name for the group R to 'sulfanyl-' (see R-3.2.1.1, Table 5), 'selanyl-', or 'tellanyl-', respectively[70].

Examples:

CH$_3$–S–
Methylsulfanyl
(traditionally Methylthio)

C$_6$H$_5$–Se–
Phenylselanyl
(traditionally Phenylseleno)

Divalent groups, such as –O–Y–O– and –S–Y–S–, are named by adding 'dioxy' or 'bis(sulfanediyl)' to the name of the divalent group –Y– .

---

[69] The prefixes 'mercapto-' and 'hydroseleno-', used in the 1979 edition of the IUPAC *Nomenclature of Organic Chemistry*[1] to designate the HS– and HSe– groups, respectively, are not included in these recommendations.
[70] In previous editions of the IUPAC *Nomenclature of Organic Chemistry*[1], substituent prefix names for RS– and RSe– were formed by adding 'thio' or 'seleno' to the substituent prefix name for the group R.

Examples:

-O-CH₂-O-   -S-SO₂-S-
Methylenedioxy   Sulfonylbis(sulfanediyl)

R-5.5.3   Salts

Anions derived from alcohols, phenols and their chalcogen analogues by loss of the hydrogen atom of the chalcogen atom as a hydron[71] are named by changing the final 'ol' of the name to 'olate'. When the group RO– has an abbreviated name, for example, methoxy, the anion name may be formed by changing the ending '-oxy' to '-oxide'.

Examples:

$CH_3-O^-Na^+$ $CH_3-CH_2-S^-Na^+$
Sodium methanolate   Sodium ethanethiolate
Sodium methoxide   Sodium ethyl sulfide

$(CH_3-CH_2-CH_2-O^-)_2Mg^{2+}$
Magnesium bis(propan-1-olate)
Magnesium dipropoxide

R-5.5.4   Ethers and chalcogen analogues

Compounds having the general structure R–O–R′, R–S–R′, R–Se–R′ and R–Te–R′ are called generically 'ethers', 'sulfides', 'selenides', and 'tellurides', respectively, and are named by one of three methods: substitutive, functional class, or replacement.

R-5.5.4.1   *Substitutive names* are formed by prefixing the name of the group R′O–, R′S–, R′Se–, or R′Te– to the name of the parent hydride corresponding to R[72].

Examples:

$CH_3-CH_2-O-CH_2-CH_2-Cl$
1-Chloro-2-ethoxyethane

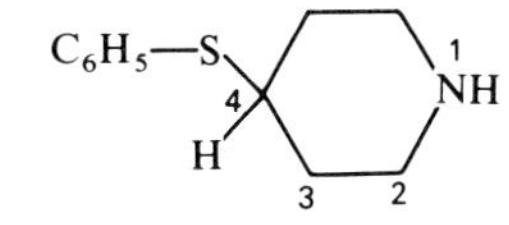

4-(Phenylsulfanyl)piperidine

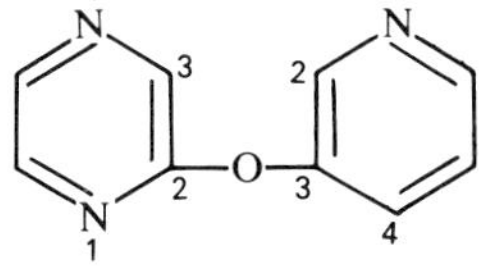

2-(3-Pyridyloxy)pyrazine

---

[71] According to the recommendations of the IUPAC Commission on Physical Organic Chemistry ['Names for Hydrogen Atoms, Ions, and Groups, and for Reactions Involving Them', *Pure Appl. Chem.*, **60**, 1115–1116 (1988)], the word 'proton' should be restricted to the cation $^1H^+$, the cation H^+ in its natural isotopic abundance being named 'hydron'.

[72] These compounds can also be named as derivatives of the appropriate parent hydrides 'oxidane', 'sulfane', 'selane', or 'tellane' (see R-2.1, Table 2).

Examples:

$CH_3-O-CH_2-CH_3$ $CH_3-CH_2-S-CH_2-CH_2-CH_3$
Ethyl(methyl)oxidane   Ethyl(propyl)sulfane

R-5.5.4.2   *Functional class names* are formed by citing the names of the groups R and R′ in alphabetical order followed by the class name 'ether', 'sulfide', 'selenide', or 'telluride', each as a separate word.

Examples:

$CH_3-O-CH_2-CH_3$
Ethyl methyl ether

$CH_3-S-CH_2-CH_2-CH_3$
Methyl propyl sulfide

$CH_3-CH_2-O-CH=CH_2$
Ethyl vinyl ether

$CH_3-CH_2-Se-CH_2-CH_3$
Diethyl selenide

Pyrazin-2-yl 3-pyridyl ether

R-5.5.4.3   *Replacement nomenclature* (see R-1.2.2) is used to name linear polyethers, polysulfides, etc. It is especially advantageous for unsymmetrical structures with several chalcogen atoms and those with several different chalcogen atoms.

Examples:

$CH_3-[CH_2]_{11}-S-[CH_2]_2-O-[CH_2]_2-O-[CH_2]_2-O-[CH_2]_2-O-CH_2-CH_2-OH$

3,6,9,12,15-Pentaoxa-18-thiatriacontan-1-ol

$$CH_3-S-CH_2-S-\overset{\overset{\displaystyle CH_3}{|}}{C}H-CH_2-S-CH_2-S-CH_2-S-CH_3$$

8-Methyl-2,4,6,9,11-pentathiadodecane

$$CH_3-CH_2-CH_2-Se-\overset{\overset{\displaystyle CH_3}{|}}{C}H-CH_2-Se-\overset{\overset{\displaystyle CH_3}{|}}{C}H-CH_3$$

2,5-Dimethyl-3,6-diselenanonane

R-5.5.4.4   *Cyclic ethers.* An oxygen atom directly attached to two atoms that are already part of a ring system, or to two carbon atoms of a chain, may be named (a) as a heterocycle, following the appropriate recommendations for naming heterocycles, including the use of the prefix 'epoxy' as a bridge prefix, in which case it is a nondetachable prefix (see R-0.1.8); or (b) by using the prefix 'epoxy' substitutively, in which case it is detachable (see R-0.1.8) and therefore alphabetized along with any other substitutive prefixes. The latter method is particularly useful when it is desired to preserve the name of a specific structure, for example, in naming natural products such as steroids and carotenoids.

Note: The class name 'oxide' has also been used additively to name cyclic ethers of the second kind, for example, ethylene oxide and styrene oxide.

Examples:

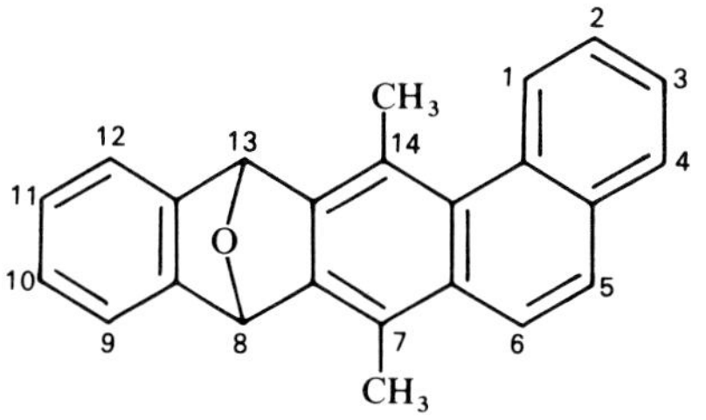

(a) 7,14-Dimethyl-8,13-dihydro-8,13-epoxybenzo[*a*]tetracene

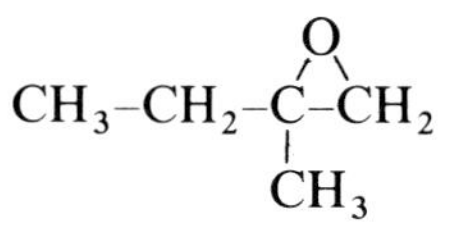

(a) 2-Ethyl-2-methyloxirane
(b) 1,2-Epoxy-2-methylbutane

R-5.5.5   **Hydroperoxides and peroxides**
Compounds with the general structure RO–OH are called generically 'hydroperoxides' and are named substitutively by citing the prefix 'hydroperoxy-' before the name of the parent hydride corresponding to R, and by functional class nomenclature by citing the prefix name of the group R followed by the class name 'hydroperoxide' as a separate word[73].

Examples:

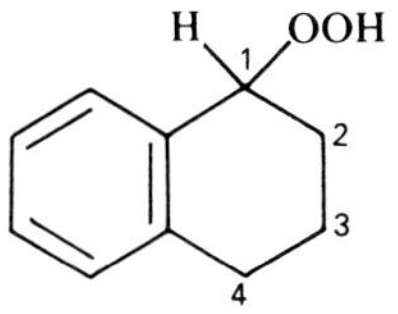

1-Hydroperoxy-1,2,3,4-tetrahydronaphthalene
1,2,3,4-Tetrahydro-1-naphthyl hydroperoxide

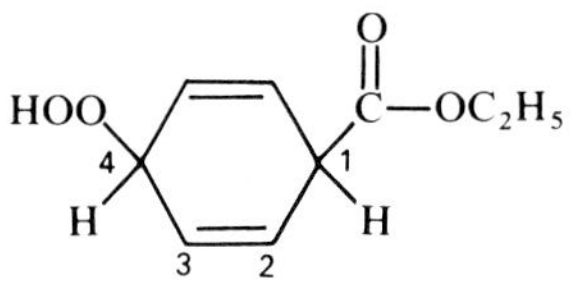

Ethyl 4-hydroperoxycyclohexa-2,5-diene-1-carboxylate

Compounds with the general structure RO–OR' are called generically 'peroxides' and are named substitutively by citing the prefix 'R'-peroxy'-[74] before the name of the parent hydride corresponding to R and by functional class nomenclature by citing the prefix

[73] Hydroperoxides may also be named as derivatives of the parent hydride 'dioxidane', HO–OH; for example, C_6H_5–O–OH would be called phenyldioxidane.
[74] In the 1979 edition of the IUPAC *Nomenclature of Organic Chemistry*[1], the prefix name for the –OO– group was 'dioxy-'.

96

names of the groups R and R', in alphabetical order, followed by the class name 'peroxide' each as a separate word[75].

Examples:

$C_6H_5-O-O-C_2H_5$
(Ethylperoxy)benzene
Ethyl phenyl peroxide

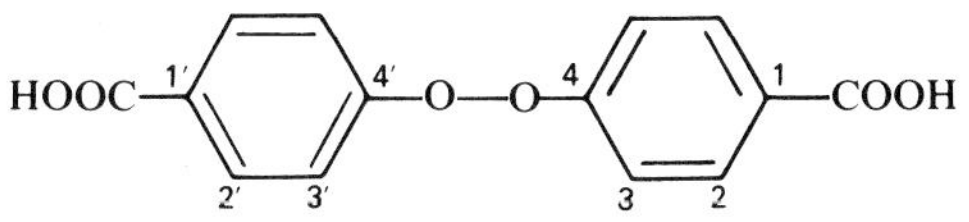

$HOOC - \langle \text{ring } 1', 2', 3', 4' \rangle - O - O - \langle \text{ring } 4, 3, 2, 1 \rangle - COOH$

4,4'-Peroxydibenzoic acid

R-5.5.6 Hydropolysulfides and polysufides

Compounds with the general structures $R-[S]_n-H$ and $R-[S]_n-R'$ are called generically 'hydropolysulfides' and 'polysulfides', respectively. They are named substitutively as derivatives of polysulfanes $H-[S]_n-H$, if the sulfur chain is linear (see R-2.2.2), and by functional class nomenclature by citing the prefix name of the group R, or the names of the groups R and R', in alphabetical order, followed by the class name 'hydropolysulfide' or 'polysulfide' respectively, each as a separate word. Selenium and tellurium analogues are named in the same way using the parent hydride names 'polyselane' or 'polytellane' and the class names 'hydropolyselenide' or 'hydropolytelluride' and 'polyselenide' or 'polytelluride', respectively.

Examples:

$C_2H_5-S-S-H$ $C_6H_5-S-S-C_6H_5$
Ethyldisulfane   Diphenyldisulfane
Ethyl hydrodisulfide   Diphenyl disulfide

$C_6H_5-[Se]_3-H$ $CH_3-CH_2-CH_2-[Se]_3-CH_3$
Phenyltriselane   Methyl(propyl)triselane
Phenyl hydrotriselenide   Methyl propyl triselenide

R-5.5.7 Sulfoxides, sulfones, and their analogues

Compounds with the general structures $R-SO-R'$ and $R-SO_2-R'$ are called generically 'sulfoxides' and 'sulfones', respectively. They may be named substitutively by citing the prefix name of the group R' followed by 'sulfinyl' or 'sulfonyl' and the name of the parent hydride corresponding to R. In functional class nomenclature, names are formed by citing the prefix names for the groups R and R' in alphabetical order as separate words followed by the class name 'sulfoxide' or 'sulfone', respectively[76]. Selenium and tellurium analogues are named in the same way using prefix names such as 'R'-seleninyl-' and 'R'-selenonyl-', and class names such as 'selenoxide' and 'selenone'.

---

[75] Peroxides may also be named as derivatives of the parent hydride 'dioxidane', HO–OH; for example, $C_6H_5-O-O-C_2H_5$ would be called ethyl(phenyl)dioxidane.

[76] Sulfoxides and sulfones can also be named substitutively by specifying the oxygen atoms by means of 'oxo-' prefixes to the parent hydride names λ^4- and λ^6-sulfane; for example, $(C_6H_5)_2SO_2$ would be called dioxodiphenyl-λ^6-sulfane.

Examples:

C₆H₅–SO–C₆H₅
(Phenylsulfinyl)benzene
Diphenyl sulfoxide

CH₃–CH₂–SO–CH₂–CH₂–CH₂–CH₃
1-(Ethylsulfinyl)butane
Butyl ethyl sulfoxide

C₆H₅–SeO–CH₂–CH₃
(Ethylseleninyl)benzene
Ethyl phenyl selenoxide

C₂H₅–SO₂–C₂H₅
(Ethylsulfonyl)ethane
Diethyl sulfone

C₆H₅–SeO₂–C₆H₅
(Phenylselenonyl)benzene
Diphenyl selenone

7-(Phenylselenonyl)quinoline
Phenyl 7-quinolyl selenone

When a group >SO or >SO₂ is part of a ring system, the oxygen atom(s) can be expressed additively by citing the class name 'oxide' after the name of the heterocycle, or substitutively by adding 'oxo-' prefixes to the name of the heterocycle in which the sulfur atoms are designated as λ^4 or λ^6.

Examples:

Thiophene 1-oxide
1-Oxo-1λ^4-thiophene

Thianthrene 5,5-dioxide
5,5-Dioxo-5λ^6-thianthrene

R-5.6   ALDEHYDES, KETONES, THEIR DERIVATIVES AND ANALOGUES

R-5.6.1 **Aldehydes, thioaldehydes, and their analogues**
The generic term 'aldehyde' refers to compounds containing a –CHO group attached to a carbon atom. Aldehydes corresponding to carboxylic acids with trivial names (see R-9.1, Table 28(a), p. 175) are named by changing the '-ic acid' or '-oic acid' ending of the acid name to '-aldehyde'. Acyclic mono- and dialdehydes are named by adding the suffixes '-al' or '-dial' to the name of the acyclic hydrocarbon with the same number of carbon atoms, eliding the final 'e' of the hydrocarbon name before 'a'. Other aldehydes are named by adding the suffix '-carbaldehyde' to the name of a parent hydride. When a group having priority for citation as a principal characteristic group is present, an aldehyde group is described by the prefix 'formyl-'. In names of natural products, conversion of an implied CH₃ group to an aldehyde is indicated by the prefix 'oxo-'.

98

Examples:

CH$_3$–CHO
Acetaldehyde

CH$_3$–CH$_2$–CH$_2$–CH$_2$–CHO
Pentanal

OHC–[CH$_2$]$_4$–CHO
Hexanedial

CHO
H
Cyclohexanecarbaldehyde

CHO
OHC–CH$_2$–CH$_2$–CH–CH$_2$–CHO
  4   3   2   1
Butane-1,2,4-tricarbaldehyde

CH$_2$–CHO
OHC–CH$_2$–CH$_2$–CH–CH$_2$–CHO
  6   5   4   3   2   1
3-(Formylmethyl)hexanedial

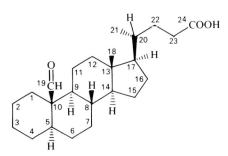
19-Oxo-5α-cholan-24-oic acid

Chalcogen analogues of aldehydes are named using suffixes such as '-thial', '-selenal', '-carbothialdehyde', and '-carboselenaldehyde' and prefixes such as 'thioformyl-' and 'thioxo-'[77]. The use of the prefixes 'thio-', 'seleno-', and 'telluro-' with trivial names of aldehydes such as acetaldehyde to indicate replacement of an aldehydic oxygen atom is not included in these recommendations.

Examples:

CH$_3$–CHS
Ethanethial

CH$_3$–[CH$_2$]$_4$–CHSe
Hexaneselenal

SHC–CH$_2$–CH$_2$–CH$_2$–CHS
Pentanedithial

C$_6$H$_5$CHS
Benzenecarbothialdehyde

SHC—⬡—COOH
    4   1
      3 2
4-(Thioformyl)benzoic acid

SeHC⬡COOH
   4   1
 H 3 2 H
4-(Selenoformyl)cyclohexane-1-carboxylic acid

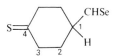
4-Thioxocyclohexane-1-carboselenaldehyde (see also R-5.6.2.2)

---

[77] The suffixes '-thiocarbaldehyde' and '-selenocarbaldehyde' have been used in the literature.

R-5.6.2   **Ketones, thioketones, and their analogues**

R-5.6.2.1   *Ketones.* The generic term 'ketone' refers to compounds containing a carbonyl group, >C=O, joined to two carbon atoms. Ketones are named substitutively by adding a suffix such as '-one', and '-dione' to the name of a parent hydride with elision of the final 'e' of the parent hydride, if any, before 'o'. When a group having priority for citation as principal characteristic group is present, a ketone is described by the prefix 'oxo-'. Functional class names for monoketones and vicinal diketones, etc., are formed by citing the prefix names for the two groups attached to the carbonyl group(s) in alphabetical order followed by the class name 'ketone', 'diketone', etc., as a separate word.

Examples:

$$CH_3-CH_2-\overset{\overset{\text{O}}{\|}}{C}-CH_3$$
$$\;\;\;4\quad\;\;\;3\quad\;\;\;2\quad\;1$$

Butan-2-one
Ethyl methyl ketone

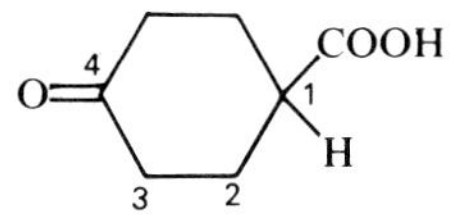

4-Oxocyclohexane-1-carboxylic acid

$$CH_3-CH_2-CH_2-CH_2-CH_2-\overset{\overset{\text{O}}{\|}}{C}-CH_3$$
$$\;\;7\quad\;\;\;6\quad\;\;\;5\quad\;\;\;4\quad\;\;\;3\quad\;\;\;2\quad\;1$$

Heptan-2-one
Methyl pentyl ketone

$$C_6H_5-CH_2-\overset{\overset{\text{O}}{\|}}{C}-\overset{\overset{\text{O}}{\|}}{C}-CH_2-CH_3$$

1-Phenylpentane-2,3-dione
Benzyl ethyl diketone

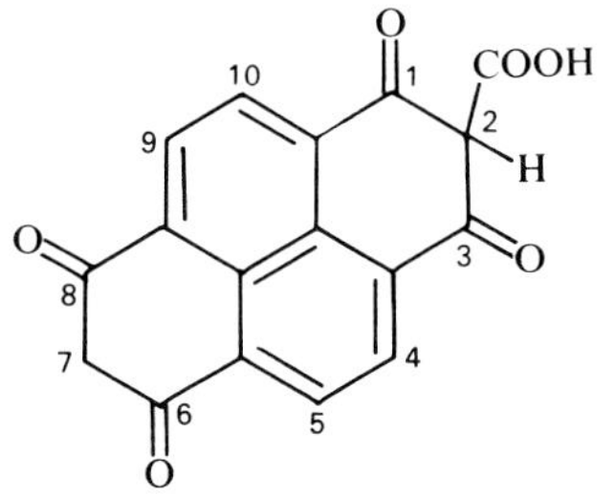

1,3,6,8-Tetraoxo-1,2,3,6,7,8-hexahydropyrene-2-carboxylic acid

$$CH_3-CH_2-\overset{\overset{\text{O}}{\|}}{C}-CH_2-\overset{\overset{\text{O}}{\|}}{C}-CH_3$$
$$\;\;6\quad\;\;\;5\quad\;\;\;4\quad\;3\quad\;\;\;2\quad\;1$$

Hexane-2,4-dione

Diketones derived from cyclic parent hydrides having the maximum number of non-cumulative double bonds by conversion of two –CH= groups into >CO groups with rearrangement of double bonds to a quinonoid structure may be named alternatively by adding the suffix '-quinone' to the name of the aromatic parent hydride.

Example:

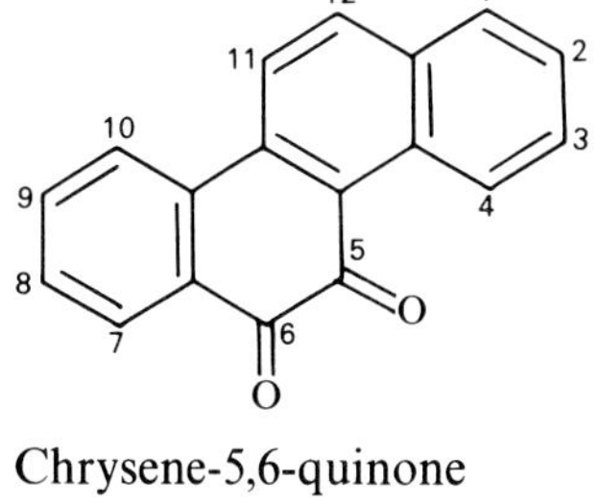

Chrysene-5,6-quinone
Chrysene-5,6-dione

100

Acyl derivatives of benzene or naphthalene have been named by changing the '-ic acid' or '-oic acid' ending of a trivial name of the acid corresponding to the acyl group to '-ophenone' or '-onaphthone'. Only the names acetophenone, propiophenone, and benzophenone are retained in these recommendations (see R-9.1, Table 27(a), p. 174). Acyl derivatives of cyclic parent hydrides are named by prefixing the substituent name derived from the cyclic parent hydride to the name of the acyclic ketone.

Example:

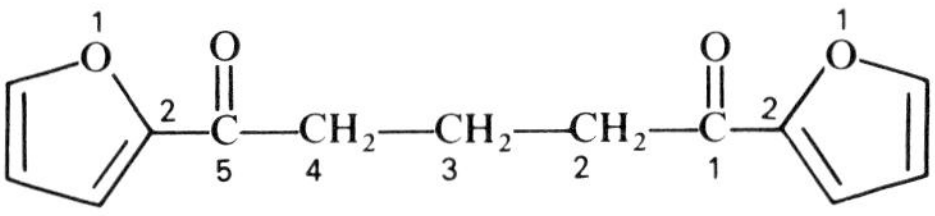

1,5-Di(2-furyl)pentane-1,5-dione

Some trivial names are retained (see R-9.1, Table 27(a), p. 174).

R-5.6.2.2   ***Chalcogen analogues of ketones*** are named by using suffixes such as '-thione' and '-selone', and prefix names such as 'thioxo-' and 'selenoxo-'. The use of prefixes such as 'thio-' and 'seleno-' with trivial names of ketones, such as acetone, to indicate replacement of the ketonic oxygen atom with a chalcogen atom is not recommended.

Examples:

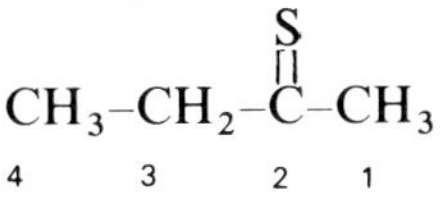

$$CH_3-CH_2-\overset{\overset{S}{\|}}{C}-CH_3$$
4   3   2   1

Butane-2-thione

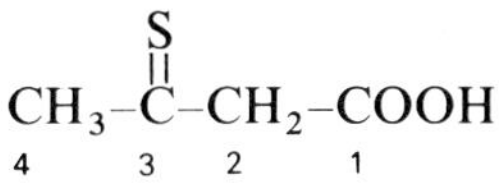

$$CH_3-CH_2-CH_2-\overset{\overset{Se}{\|}}{C}-CH_2-CH_3$$
6   5   4   3   2   1

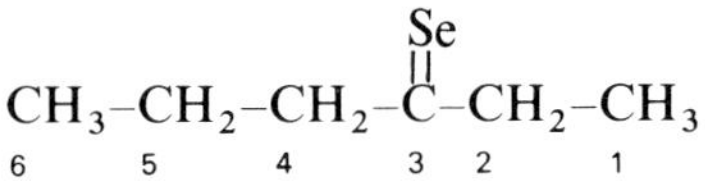

Hexane-3-selone

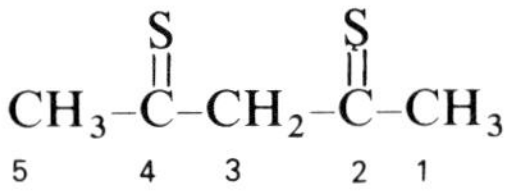

$$CH_3-\overset{\overset{S}{\|}}{C}-CH_2-\overset{\overset{S}{\|}}{C}-CH_3$$
5   4   3   2   1

Pentane-2,4-dithione

$$CH_3-\overset{\overset{S}{\|}}{C}-CH_2-COOH$$
4   3   2   1

3-Thioxobutanoic acid

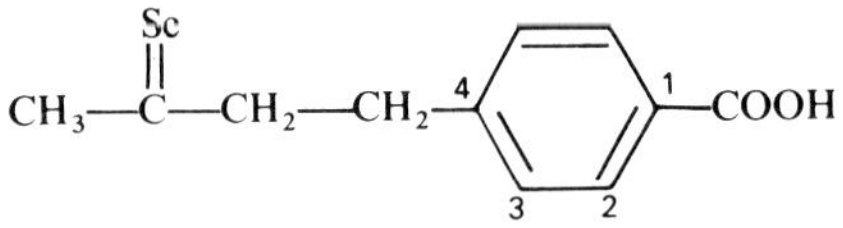

4-(3-Selenoxobutyl)benzoic acid

R-5.6.3   **Ketenes**
The compound $CH_2=C=O$ is a functional parent compound with the name 'ketene'. Derivatives can be named by (a) citing all substituents as prefixes or (b) using the principles for naming ketones (see R-5.6.2.1).

Examples:

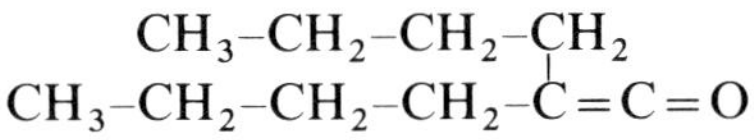

$$CH_3-CH_2-CH_2-CH_2$$
$$CH_3-CH_2-CH_2-CH_2-C=C=O$$
(a) Dibutylketene
(b) 2-Butylhex-1-en-1-one

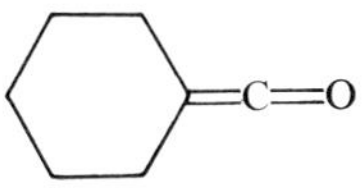

(b) Cyclohexylidenemethanone

R-5.6.4   **Acetals, hemiacetals, acylals, and their analogues**

R-5.6.4.1   *Acetals.* Compounds with the general structure RR'C(O–R")(O–R'''), where R and/or R' may be, but need not be, hydrogen, but R" and R''' cannot be hydrogen, are termed generically 'acetals'. 'Ketals' constitute a subclass of acetals wherein neither R nor R' may be hydrogen[78]. Acetals are named substitutively as 'alkoxy-', 'aryloxy-', etc., derivatives of an appropriate parent hydride or functional parent compound. Alternatively, according to functional class nomenclature, the name of the appropriate aldehyde or ketone is followed by that of the *O*-substituent(s), which is in turn followed, after a space, by the word 'acetal' or 'ketal'.

Examples:

$$O\text{–}CH_2\text{–}CH_3$$
$$CH_3\text{–}CH_2\text{–}\underset{3\quad\ 2\quad\ 1}{CH}\text{–}O\text{–}CH_2\text{–}CH_3$$

1,1-Diethoxypropane
Propanal diethyl acetal
Propionaldehyde diethyl acetal

1-Ethoxy-1-methoxycyclohexane
Cyclohexanone ethyl methyl ketal

Sulfur analogues of acetals and ketals with the general structures RR'C(S–R")(S–R''') or RR'C(O–R")(S–R'''), are termed generically 'dithioacetals' or 'monothioacetals', respectively. They are named substitutively as 'alkylsulfanyl-', 'arylsulfanyl-', 'alkyloxy-' (or 'alkoxy-'), or 'aryloxy-', derivatives, as appropriate, of a parent hydride. They may also be named by functional class nomenclature in the same way as acetals (see above). Capital italic letter locants are used to provide structural specificity. Selenium and tellurium and mixed analogues are treated in the same way as their sulfur analogues; generically, they are 'monoselenoacetals', 'ditelluroacetals', 'selenothioacetals', etc., and are named substitutively using prefixes such as 'alkylselanyl-', 'aryltellanyl-', etc.

Examples:

$$S\text{–}CH_2\text{–}CH_3$$
$$CH_3\text{–}CH_2\text{–}CH_2\text{–}CH_2\text{–}\underset{5\quad\ 4\quad\ 3\quad\ 2\quad\ 1}{CH}\text{–}S\text{–}CH_2\text{–}CH_3$$

1,1-Bis(ethylsulfanyl)pentane
Pentanal diethyl dithioacetal

$$O\text{–}CH_3$$
$$CH_3\text{–}CH_2\text{–}\underset{3\quad\ 2\quad\ 1}{CH}\text{–}S\text{–}CH_2\text{–}CH_3$$

1-(Ethylsulfanyl)-1-methoxypropane
Propanal *S*-ethyl *O*-methyl
monothioacetal

$$S\text{–}CH_2\text{–}CH_3$$
$$CH_3\text{–}CH_2\text{–}CH_2\text{–}\underset{4\quad\ 3\quad\ 2\quad\ 1}{CH}\text{–}O\text{–}CH_2\text{–}CH_3$$

1-Ethoxy-1-(ethylsulfanyl)butane
Butanal diethyl monothioacetal

1-(Ethylselanyl)-1-(methylsulfanyl)cyclopentane
Cyclopentanone *Se*-ethyl *S*-methyl
selenothioketal

---

[78] The term 'ketal' was abandoned in the previous rules. However, because of its continued popularity, it is reintroduced in the present document as a subclass of acetals and as a functional class term. The same applies to the cognate term 'hemiketal' (see R-5.6.4.2).

R-5.6.4.2   *Hemiacetals.* Compounds with the general structure $RR'C(OH)(O-R'')$ are termed generically 'hemiacetals' (see footnote to R-5.6.4.1). Hemiacetals are named substitutively as alkoxy-, aryloxy-, etc., derivatives of an appropriate hydroxy parent compound, such as an alcohol (see R-5.5.1.1), and by functional class nomenclature in the same way as acetals (see R-5.6.4.1) using the class name 'hemiacetal'.

Example:

$$CH_3-CH_2-O$$
$$CH_3-CH_2-CH_2-\overset{|}{C}H-OH$$
$$4321$$

1-Ethoxybutan-1-ol
Butanal ethyl hemiacetal

Sulfur analogues of hemiacetals with general structures $RR'C(SH)(S-R'')$ and $RR'C(OH)(S-R'')$ or $RR'C(SH)(O-R'')$ are generically 'dithiohemiacetals' or 'monothiohemiacetals', respectively. They are named substitutively as 'alkylsulfanyl-', 'arylsulfanyl-', 'alkyloxy-' (or 'alkoxy-'), or 'aryloxy-', derivatives, as appropriate, of a hydroxy parent compound, such as an alcohol (see R-5.5.1.1), or of a thiol parent compound (see R-5.5.1.2). They may also be named by functional class nomenclature in the same way as hemiacetals (see above, R-5.6.4.2). Selenium and tellurium and mixed analogues are treated in the same way as their sulfur analogues; generically, they are 'monoselenohemiacetals', 'ditellurohemiacetals', 'selenothiohemiacetals', etc., and are named substitutively as derivatives of a hydroxy parent compound or a chalcogen analogue, such as a 'thiol' or 'selenol', using prefixes such as 'alkylselanyl-' and 'aryltellanyl-', etc.

Examples:

$$CH_3-CH_2-S$$
$$CH_3-CH_2-\overset{|}{C}H-SH$$
$$321$$

1-(Ethylsulfanyl)propane-1-thiol
Propanal ethyl dithiohemiacetal

1-(Ethylsulfanyl)cyclohexane-1-selenol
Cyclohexanone S-ethyl selenothiohemiketal

$$CH_3-CH_2-O$$
$$CH_3-CH_2-\overset{|}{C}H-SH$$
$$321$$

1-Ethoxypropane-1-thiol
Propanal O-ethyl monothiohemiacetal

R-5.6.4.3   *Acylals.* Compounds with the general structure $R-CH(OCO-R')_2$, $RR'C(OCO-R'')_2$, etc., are generically called 'acylals'. Specific compounds are named as esters.

Example:

$$O-CO-CH_2-CH_3$$
$$CH_3-\overset{|}{C}H-O-CO-CH_2-CH_3$$
$$21$$

Ethane-1,1-diyl dipropionate
(traditionally Ethylidene dipropionate,
    see Note under R-2.5)

R-5.6.5   **Acyloins**

α-Hydroxy ketones, RCH(OH)–CO–R, in which R is an alkyl, aryl, or heterocyclic group, are generically called 'acyloins' and are named by substitutive nomenclature. Names ending in '-oin' are not included in these recommendations.

Examples:

3-Hydroxybutan-2-one

1,2-Di(2-furyl)-2-hydroxyethan-1-one

2-Hydroxy-1,2-diphenylethan-1-one

R-5.6.6   **Nitrogenous derivatives of carbonyl compounds[79]**

R-5.6.6.1   *Oximes.* Compounds having the general structure R–CH=N–OH or RR'C=N–OH are called generically 'oximes' and have been further classified as 'aldoximes' and 'ketoximes', respectively. They are named according to the principles of functional class nomenclature by placing the class name 'oxime' as a separate word after the name of the aldehyde RCHO or the ketone RR'C=O, respectively; or substitutively by use of the prefix 'hydroxyimino-' attached to the name of a parent hydride or parent substituent prefix. Compounds containing the group =N–OR may be named as O-substituted oximes or as alkoxy-substituted imines.

Examples:

C_6H_5–CH=CH–CH=N–OH
Cinnamaldehyde oxime

4-(Hydroxyimino)-1-methylcyclohexa-2,5-diene-1-carboxylic acid[81]

Pentan-3-one oxime

CH_3–CH_2–CH=N–O–C_2H_5

Propanal O-ethyloxime
N-Ethoxypropan-1-imine

C_6H_5–CH=N–OH
Benzaldehyde oxime[80]

4-[(Ethoxyimino)methyl]benzene-1-sulfonic acid

[79] Imines have been considered as derivatives of carbonyl compounds, for example, p-benzoquinone diimine; however, the recommended nomenclature for imines is described in R-5.4.3.
[80] The contracted form benzaldoxime has been used.
[81] For indexing purposes, it may be convenient to designate a functional modification of a characteristic group described in a name by a prefix, such as 'oxo-', by a functional class name, such as oxime; such a name for this example would be 1-methyl-4-oxocyclohexa-2,5-diene-1-carboxylic acid 4-oxime.

104

R-5.6.6.2   **Hydrazones.** Compounds having the general structure $RCH=N-NH_2$ or $RR'C=N-NH_2$ are called 'hydrazones' generically and are named according to functional class nomenclature by placing the class name 'hydrazone' as a separate word following the name of the corresponding aldehyde or ketone, or substitutively by means of the prefix 'hydrazono-' attached to the name of an appropriate parent hydride or parent substitutive prefix.

Examples:

$CH_3-CH_2-CH=N-NH_2$
Propanal hydrazone

4-Hydrazonocyclohexane-1-carboxylic acid

R-5.6.6.3   **Azines.** Compounds having the general structure $RCH=N-N=CHR$ or $RR'C=N-N=CRR'$ are called 'azines' generically and are named substitutively as derivatives of diazane or as an assembly of identical units using the prefix 'azino-'. According to functional class nomenclature, they may be named by adding the class name 'azine' as a separate word after the name of the corresponding aldehyde or ketone.

Examples:

$(CH_3)_2C=N-N=C(CH_3)_2$

1,2-Diisopropylidenediazane
Acetone azine

4,4'-Azinodi(cyclohexane-1-carboxylic acid)

Compounds having the general structure $R=N-N=R'$ ($R \neq R'$) are named as 'ylidenehydrazones' (see R-5.6.6.2) of the preferred carbonyl compound or by using the prefix 'hydrazono-'.

Examples:

Cyclohexanone isopropylidenehydrazone

4-(Isopropylidenehydrazono)cyclohexa-2,5-diene-1-carboxylic acid

R-5.6.6.4   **Other nitrogen derivatives of carbonyl compounds.** Derivatives of semicarbazide

$$NH_2-CO-NH-NH_2$$
(numbered 4, 3, 2, 1)

of the types $NH_2-CO-NH-N=CH-R$ and $NH_2-CO-NH-N=CRR'$ may be named substitutively by using the functional parent compound name 'semicarbazide', or the prefix name 'semicarbazono-', or according to principles of functional class nomenclature, by adding the class name 'semicarbazone' as a separate word after the name of the aldehyde RCHO or ketone $RR'C=O$. Chalcogen analogues are named in the same way on the basis of functional parent compound names such as 'selenosemicarbazide' and prefix names such as 'thiosemicarbazono-', or functional class names such as 'thiosemicarbazone'.

Examples:

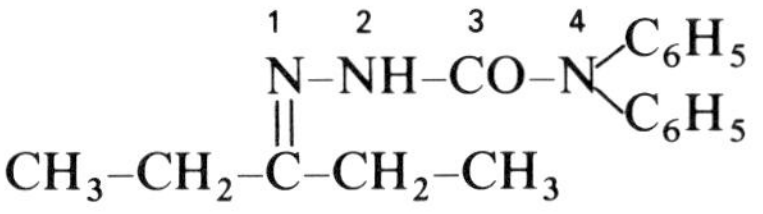

$$\underset{CH_3-CH_2-\overset{\|}{C}-CH_2-CH_3}{\overset{1\quad 2\quad 3\quad 4}{N-NH-CO-N\overset{C_6H_5}{\underset{C_6H_5}{<}}}}$$

1-(Pentan-3-ylidene)-4,4-diphenylsemicarbazide
1-(1-Ethylpropylidene)-4,4-diphenylsemicarbazide
Pentan-3-one 4,4-diphenylsemicarbazone

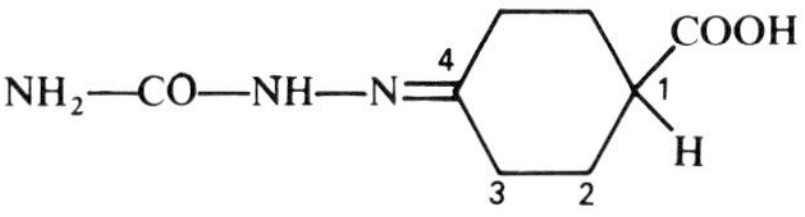

$$NH_2-CO-NH-N=$$

4-Semicarbazonocyclohexane-1-carboxylic acid
4-Oxocyclohexane-1-carboxylic acid semicarbazone

$$\underset{CH_3-CH_2-\overset{\|}{C}-CH_3}{N-NH-CSe-NH_2}$$

1-(Butan-2-ylidene)selenosemicarbazide
1-(1-Methylpropylidene)selenosemicarbazide
Butan-2-one selenosemicarbazone

Other derivatives of semicarbazide, and derivatives of carbonohydrazide, $NH_2-NH-CO-NH-NH_2$, carbazone, $NH=N-CO-NH-NH_2$, and their chalcogen analogues may be named substitutively using these functional parent compound names and prefix names such as 'semicarbazido-', 'thiocarbonohydrazido-' and 'carbazono-'. These prefix names describe a substituent derived from the appropriate functional parent compound names by loss of one hydrogen atom from position '1' only.

Examples:

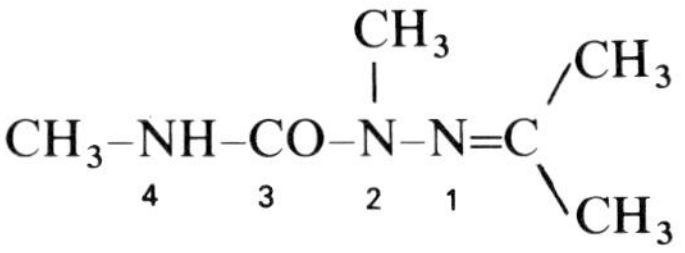

$$\underset{4\quad 3\quad 2\quad 1}{CH_3-NH-CO-N-N=\overset{\overset{CH_3}{|}}{C}\overset{/CH_3}{\underset{\backslash CH_3}{}}}$$

1-Propan-2-ylidene-2,4-dimethylsemicarbazide
1-Isopropylidene-2,4-dimethylsemicarbazide
Acetone 2,4-dimethylsemicarbazone

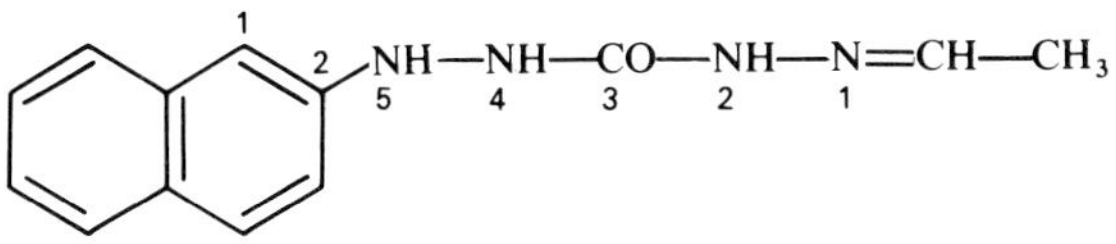

$$\underset{5\quad 4\quad 3\quad 2\quad 1}{NH-NH-CO-NH-N=CH-CH_3}$$

1-Ethylidene-5-(2-naphthyl)carbonohydrazide

R-5.7   ACIDS AND RELATED CHARACTERISTIC GROUPS

Carboxylic acids and related characteristic groups, except for aldehydes, for which see R-5.6.1, are named by means of an appropriate suffix or ending as listed in Table 12.

Table 12 Suffixes and endings for carboxylic acids, some related characteristic groups, and substituted derivatives

(a) *Monocarboxylic acids and related characteristic groups*

–(C)OOH	-oic acid	–COOH	-carboxylic acid
–(C)OOR	R … oate	–COOR	R … carboxylate
⌒(C)O–O⌒	–olactone	⌒CO–O⌒	-carbolactone
–(C)O–X	-oyl halide	–CO–X	-carbonyl halide
[–(C)O]$_2$O	-oic anhydride	–(CO)$_2$O	-carboxylic anhydride
–(C)O–NH$_2$	-amide	–CONH$_2$	-carboxamide
–(C)≡N	-nitrile	–C≡N	-carbonitrile

(b) *Dicarboxylic acids, related characteristic groups, and substituted derivatives*

HOO(C)…(C)OOH	-dioic acid	HOOC…COOH	-dicarboxylic acid
⌒(C)O–NH–(C)O⌒	-imide[a]	⌒CO–NH–CO⌒	-dicarboximide
HOO(C)…(C)O–NH$_2$	-amic acid[a]		
HOO(C)…(C)O–NH–C$_6$H$_5$	-anilic acid[a]		
HOO(C)…(C)HO	-aldehydic acid[a]		

[a] This ending is used only to replace the '-ic acid' ending of a *trivial name* of a dibasic acid.

Suffixes to designate modified carboxylic acid groups in which the carbonyl oxygen atom or the hydroxy group is replaced by another atom or group are illustrated in Table 13. Organic sulfur and phosphorus acids and replacement modifications are illustrated in Tables 14 and 15, respectively.

R-5.7.1 **Carboxylic acids**

R-5.7.1.1 ***Simple (unsubstituted) acyclic mono- and dicarboxylic acids.*** Carboxylic acid groups, –COOH, that conceptually replace CH$_3$ groups terminating an acyclic hydrocarbon chain are denoted by adding the suffix '-oic acid' or '-dioic acid' (see Table 12) to the name of the acyclic hydrocarbon with elision of its final 'e' before 'o'.

Examples:

CH$_3$–[CH$_2$]$_5$–COOH   HOOC–[CH$_2$]$_8$–COOH
Heptanoic acid   Decanedioic acid

If an unbranched chain is directly linked to more than two carboxy groups, these carboxy groups are named from the parent hydride by substitutive use of a suffix such as '-tricarboxylic acid', etc.

Example:

                COOH
                |
HOOC–CH$_2$–CH$_2$–CH–CH$_2$–CH$_2$–COOH
Pentane-1,3,5-tricarboxylic acid
(**not** 4-Carboxyheptanedioic acid)

Other carboxylic acids are named by adding the suffix '-carboxylic acid' to the name of a parent hydride.

Examples:

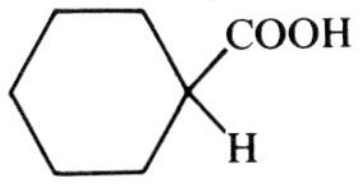

Cyclohexanecarboxylic acid

$H_3Si–SiH_2–COOH$
Disilanecarboxylic acid

When another group is present that has priority for citation as a suffix (see Table 10, R-4.1 and Table 5, R-3.2.1.1) or when all carboxylic acid groups cannot be described in the suffix, a carboxylic acid group is indicated by the prefix 'carboxy-'.

Examples:

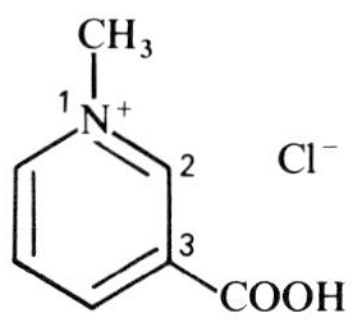

3-Carboxy-1-methylpyridinium chloride

$$\underset{7\quad\;6\qquad\;5\qquad\;4\quad\;\;3\quad\;2\qquad\;1}{HOOC–CH_2–CH_2–CH_2–\overset{\displaystyle CH_2–COOH}{CH}–CH_2–COOH}$$

3-(Carboxymethyl)heptanedioic acid

The name of a monovalent or divalent acyl group formed by removal of the –OH group from each carboxy group of a carboxylic acid denoted by an '-oic acid' suffix or having a trivial name (see Table 28, p. 175) is derived from the name of the corresponding acid by changing the ending '-oic acid' or '-ic acid' to '-oyl' or '-yl', respectively.

Examples:

$CH_3–[CH_2]_5–CO–$
Heptanoyl

$CH_3–CO–$
Acetyl

$–CO–[CH_2]_8–CO–$
Decanedioyl

$–CO–CH_2–CO–$
Malonyl

An acyl group derived from an acid named by means of the suffix '-carboxylic acid' is named by changing the suffix to '-carbonyl'.

Examples:

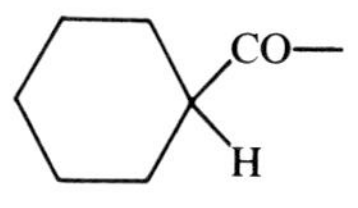

Cyclohexanecarbonyl

$$\underset{6\qquad\;5\quad\;\;4\qquad\;3\quad\;\;2\qquad1}{CH_3–\overset{\displaystyle CO}{CH}–CH_2–\overset{\displaystyle CO}{CH}–\overset{\displaystyle CO}{CH}–CH_3}$$

Hexane-2,3,5-tricarbonyl

Some trivial names are retained (see R-9.1, Table 28(a), p. 175).

R-5.7.1.2   ***Substituted carboxylic acids***

R-5.7.1.2.1   *Hydroxy, alkoxy, and oxo acids.* Some trivial names for hydroxy and alkoxy acids are retained (see Section R-9.1, Table 28(b), p. 176). The names of carboxylic acids containing an aldehydic group attached to, or a ketonic group contained in the principal chain or parent ring system, are generally derived from the names of the corresponding simple carboxylic acids by adding prefixes such as 'oxo-', 'dioxo-', etc., denoting =O substituents, or 'formyl-', denoting a –CHO substituent.

Examples:

$$\overset{\overset{\displaystyle O}{\|}}{CH_3-C}-CH_2-CH_2-CH_2-COOH$$
6   5   4   3   2   1

5-Oxohexanoic acid

$$OHC-CH_2-CH_2-\overset{\overset{\displaystyle O}{\|}}{C}-CH_2-COOH$$
6   5   4   3   2   1

3,6-Dioxohexanoic acid
5-Formyl-3-oxopentanoic acid

$$CH_3-\overset{\overset{\displaystyle O}{\|}}{C}-CH_2-\overset{\overset{\displaystyle O}{\|}}{C}-CH_2-COOH$$
6   5   4   3   2   1

3,5-Dioxohexanoic acid

4-Formyl-2-oxocyclohexane-1-carboxylic acid

When a dicarboxylic acid has a trivial name (see R-9.1, Table 28(a), p. 175), the replacement of one of the carboxy groups by an aldehydic group may be denoted by changing the ending '-ic acid' into '-aldehydic acid' (see Table 12(b)).

Examples:

OHC–CH$_2$–COOH
Malonaldehydic acid

Phthalaldehydic acid

R-5.7.1.2.2   *Amic and anilic acids.* When a dicarboxylic acid has a retained trivial name (see R-9.1, Table 28(a), p. 175) and when one of its carboxy groups is replaced by a carboxamide group –CO–NH$_2$, the resulting amic acid is named by replacing the suffix '-ic acid' of the name of the dicarboxylic acid by the suffix '-amic acid'. Some trivial names are retained (see R-9.1, Table 28(c), p. 176).

Examples:

H$_2$N–CO–CH$_2$–CH$_2$–COOH
Succinamic acid

3-Bromophthalamic acid

N-Phenyl derivatives of amic acids may be named by changing the '-amic acid' suffix to '-anilic acid'. Positions on the phenyl ring are indicated by primed numbers. The nitrogen atom is indicated by '*N*'.

109

Example:

$$C_6H_5-NH-\overset{\overset{\displaystyle O}{\displaystyle \|}}{C}-CH_2-CH_2-COOH$$

Succinanilic acid

R-5.7.1.2.3   *Amino acids*. Retained trivial names for amino carboxylic acids are given in R-9.1, Tables 28(b) and (c), p. 176. α-Amino carboxylic acids are treated in specialized rules[82].

R-5.7.1.3   ***Modification of carboxylic acid suffixes***

R-5.7.1.3.1   *Peroxy acids*. Acids containing the group –CO–OOH are called generically 'peroxy acids' and are named by placing prefixes such as 'peroxy-', 'monoperoxy-', and 'diperoxy-', as appropriate, before a trivial (see R-9.1, Table 28(a), p. 175) or systematic name of an '-oic acid', or before a '-carboxylic acid' or '-dicarboxylic acid' suffix (see Table 13). When another group is present that has priority for citation as a suffix (see Table 10, R-4.1 and Table 5, R-3.2.1.1), a peroxy carboxylic acid group is indicated by the prefix 'hydroperoxycarbonyl-'.

Examples:

$CH_3-CH_2-CO-OOH$
Peroxypropionic acid

$HOO-CO-OOH$
Diperoxycarbonic acid

$CH_3-CH_2-CH_2-CH_2-CH_2-CO-OOH$
Peroxyhexanoic acid

$HOOC-[CH_2]_4-CO-OOH$
Monoperoxyhexanedioic acid

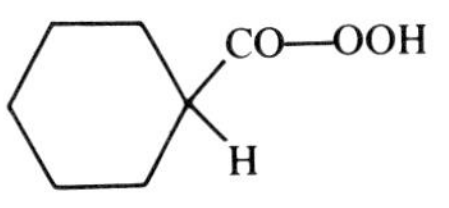

Cyclohexaneperoxycarboxylic acid

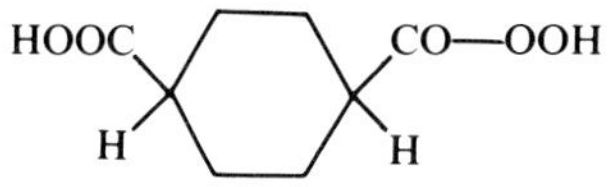

Cyclohexanemonoperoxy-1,4-dicarboxylic acid

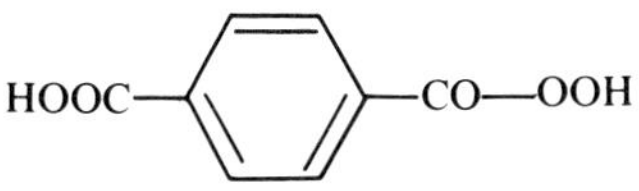

Monoperoxyterephthalic acid

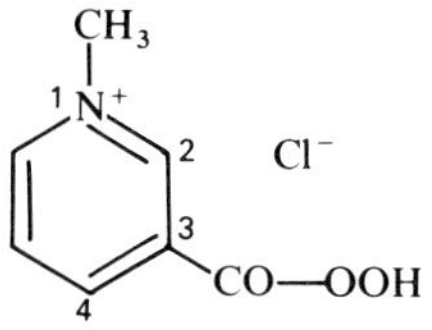

3-(Hydroperoxycarbonyl)-1-methylpyridinium chloride

R-5.7.1.3.2   *Imidic, hydrazonic, and hydroximic acids*. The name of an acid in which the carbonyl oxygen atom of a carboxylic acid group has been replaced by a =NH, =N–NH₂, or =N–OH group is formed by modifying the '-oic' or '-carboxylic' suffix of a systematic name of an acid, or the '-ic acid' ending of the trivial name of an acid to '-imidic' or

[82] International Union of Pure and Applied Chemistry and International Union of Biochemistry. Joint Commission on Biochemical Nomenclature, 'Nomenclature and Symbolism for Amino Acids and Peptides', *Pure Appl. Chem*, **56**, 595–624 (1984).

'-carboximidic acid', '-ohydrazonic' or '-carbohydrazonic acid', '-ohydroximic' or '-carbohydroximic acid', respectively (see Table 13 and R-3.4).

Examples:

$$\underset{\text{Butanimidic acid}}{CH_3-CH_2-CH_2-\overset{\overset{\displaystyle NH}{\|}}{C}-OH}$$

$$\underset{\text{Acetohydroximic acid}}{CH_3-\overset{\overset{\displaystyle N-OH}{\|}}{C}-OH}$$

Cyclohexanecarbohydrazonic acid

Table 13 Suffixes for replacement analogues of carboxylic acids

	Replacement of –OH by another O-containing group	Replacement of =O by =S and/or –OH by –SH[83]	Replacement of =O by =NH and/or –OH by –NH–
$\overset{\overset{\displaystyle O}{\|}}{-(C)-OH}$ -oic acid	$\overset{\overset{\displaystyle O}{\|}}{-(C)-OOH}$ -peroxy…oic acid	$\overset{\overset{\displaystyle S}{\|}}{-(C)-OH}$ -thioic O-acid	$\overset{\overset{\displaystyle NH}{\|}}{-(C)-OH}$ -imidic acid
		$\overset{\overset{\displaystyle O}{\|}}{-(C)-SH}$ -thioic S-acid	$\overset{\overset{\displaystyle N-NH_2}{\|}}{-(C)-OH}$ -hydrazonic acid
		$\overset{\overset{\displaystyle S}{\|}}{-(C)-SH}$ -dithoic acid	$\overset{\overset{\displaystyle N-OH}{\|}}{-(C)-OH}$ -hydroximic acid
			$\overset{\overset{\displaystyle O}{\|}}{-(C)-NH-OH}$ -hydroxamic acid[84]
$\overset{\overset{\displaystyle O}{\|}}{-C-OH}$ -carboxylic acid	$\overset{\overset{\displaystyle O}{\|}}{-C-OOH}$ -peroxycarboxylic acid	$\overset{\overset{\displaystyle S}{\|}}{-C-OH}$ -carbothioic O-acid	$\overset{\overset{\displaystyle NH}{\|}}{-C-OH}$ -carboximidic acid
		$\overset{\overset{\displaystyle O}{\|}}{-C-SH}$ -carbothioic S-acid	$\overset{\overset{\displaystyle N-NH_2}{\|}}{-C-OH}$ -carbohydrazonic acid
		$\overset{\overset{\displaystyle S}{\|}}{-C-SH}$ -carbodithioic acid	$\overset{\overset{\displaystyle N-OH}{\|}}{-C-OH}$ -carbohydroximic acid
			$\overset{\overset{\displaystyle O}{\|}}{-C-NH-OH}$ -carbohydroxamic acid[84]

[83] Selenium analogues are named using 'selenoic' in place of 'thioic'.
[84] In these recommendations, hydroxamic acid is used as a class name; specific compounds are preferably named as *N*-hydroxy amides.

R-5.7.1.3.3   *Hydroxamic acids*. The name of an acid in which the hydroxy group of the carboxy group has been replaced by a –NH–OH group can be formed by modifying the '-oic acid' or '-carboxylic acid' suffix of a systematic name of an acid, or the '-ic acid' ending of a trivial acid name to '-ohydroxamic acid' or '-carbohydroxamic acid' (see Table 13); however, in these recommendations, hydroxamic acids are preferably named as *N*-hydroxy amides.

Example:

$$\text{O}$$
$$\|$$
CH$_3$–C–NH–OH
N-Hydroxyacetamide
(traditionally Acetohydroxamic acid)

R-5.7.1.3.4   *Thiocarboxylic and thiocarbonic acids*. Replacement of oxygen atom(s) of a carboxylic acid group or of carbonic acid by another chalcogen is indicated by the affixes 'thio', 'seleno', and 'telluro'. These names do not differentiate between tautomeric forms of mixed chalcocarboxylic or chalcocarbonic acids; such nonspecificity may be shown in a formula by a structure such as:

$$-C{\begin{Bmatrix}O\\S\end{Bmatrix}}H$$

In names, tautomeric groups in mixed chalcocarboxylic and chalcocarbonic acids, such as –CS–OH and –CO–SH, may be distinguished by prefixing italic element symbols, such as *O*– or *S*–, respectively, to the term 'acid' (see Table 13); or by prefixes such as 'hydroxy(thiocarbonyl)-' and 'sulfanylcarbonyl-'.

Replacement of oxygen by (an)other chalcogen atom(s) in a carboxylic acid having a retained trivial name or in carbonic acid is indicated by prefixes, such as 'thio-', 'seleno-', 'dithio-', etc.

Examples:
CH$_3$–CS–OH
Thioacetic *O*-acid

CH$_3$–CO–SH
Thioacetic *S*-acid

C$_6$H$_5$–C${\begin{Bmatrix}Se\\O\end{Bmatrix}}$H
Selenobenzoic acid

S=C(SH)$_2$
Trithiocarbonic acid

Replacement of oxygen by (an)other chalcogen atom(s) in a carboxylic acid having a systematic name is indicated by modifying the '-oic acid' or '-carboxylic acid' suffix to suffixes such as '-thioic acid', '-selenoic acid', '-carbodithioic acid', and '-carboselenothioic acid'; and the prefix 'carboxy-' to prefixes such as thiocarboxy-', 'diselenocarboxy-', and 'selenothiocarboxy-'.

Examples:

CH$_3$–[CH$_2$]$_4$–CS–OH
Hexanethioic *O*-acid

N—CSSH
Piperidine-1-carbodithioic acid

$CH_3-[CH_2]_4-C\begin{Bmatrix}S\\Se\end{Bmatrix}H$

Hexaneselenothioic acid

$CH_3-[CH_2]_4-C(S)-SeH$
Hexaneselenothioic Se-acid

$H\begin{Bmatrix}O\\S\end{Bmatrix}C-[CH_2]_4-C\begin{Bmatrix}O\\S\end{Bmatrix}H$

Hexanebis(thioic) acid[85]

$HSSC-CH_2-CH_2-CH_2-CH_2-CSSH$
Hexanebis(dithioic) acid

$H\begin{Bmatrix}O\\S\end{Bmatrix}C-\underset{3}{CH_2}-\underset{2}{CH_2}-\underset{1}{COOH}$

3-(Thiocarboxy)propanoic acid

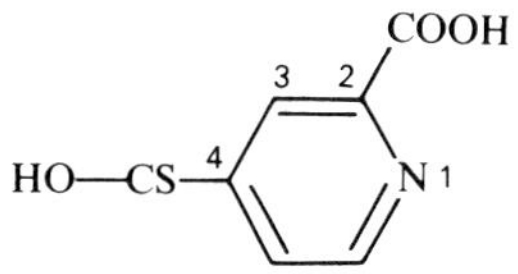

4-[Hydroxy(thiocarbonyl)]pyridine-2-carboxylic acid

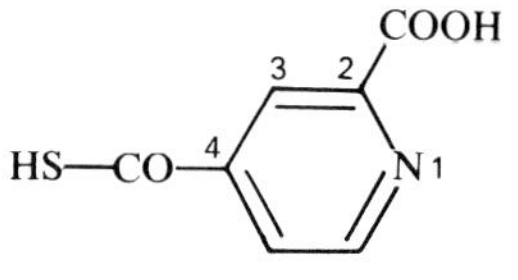

4-(Sulfanylcarbonyl)pyridine-2-carboxylic acid

$CH_3-[CH_2]_4-C\begin{Bmatrix}Se\\O\end{Bmatrix}H$

Hexaneselenoic acid

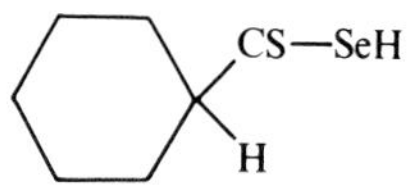

Cyclohexanecarboselenothioic Se-acid

R-5.7.2   **Chalcogen acids containing chalcogen atoms directly linked to an organic group**

R-5.7.2.1   *Sulfur acids containing sulfur atoms directly linked to an organic group* are named substitutively on the basis of a parent hydride name by adding an appropriate suffix as illustrated in Table 14. The prefixes 'sulfo-' and 'sulfino-' are used to express sulfonic and sulfinic acid groups as substituents, respectively.

Examples:

$C_6H_5-SO_2-OH$
Benzenesulfonic acid

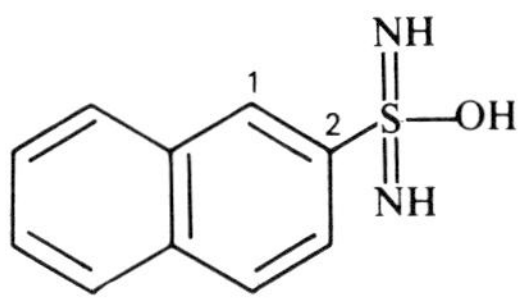

Naphthalene-2-sulfonodiimidic acid

---

[85] The multiplicative prefix 'bis-' and enclosing marks are used to avoid the ambiguity presented by the suffix '-dithioic acid', which describes a dithio analogue of a monocarboxylic acid.

Table 14 Suffixes for sulfur acids and replacement modifications

	Replacement of –OH by –SH and/or =O by =S	Replacement of =O by =NH
O ‖ –S–OH -sulfinic acid	S ‖ –S–OH -thiosulfinic *O*-acid	NH ‖ –S–OH -sulfinimidic acid
	O ‖ –S–SH -thiosulfinic *S*-acid	N–NH₂ ‖ –S–OH -sulfinohydrazonic acid
	S ‖ –S–SH -dithiosulfinic acid	N–OH ‖ –S–OH -sulfinohydroximic acid
O ‖ –S–OH ‖ O -sulfonic acid	O ‖ –S–OH ‖ S -thiosulfonic *O*-acid	NH ‖ –S–OH ‖ O -sulfonimidic acid
	O ‖ –S–SH ‖ O -thiosulfonic *S*-acid	N–NH₂ ‖ –S–OH ‖ O -sulfonohydrazonic acid
	S ‖ –S–OH ‖ S -dithiosulfonic *O*-acid	N–OH ‖ –S–OH ‖ O -sulfonohydroximic acid
	O ‖ –S–SH ‖ S -dithiosulfonic *S*-acid	NH ‖ –S–OH ‖ S -thiosulfonimidic *O*-acid
	S ‖ –S–SH ‖ S -trithiosulfonic acid	NH ‖ –S–SH ‖ O -thiosulfonimidic *S*-acid
		NH ‖ –S–OH ‖ NH -sulfonodiimidic acid
–S(O)₂–O– -sultone		–S(O)₂–NH– -sultam

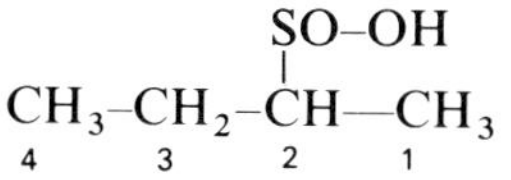

SO–OH
|
CH_3–CH_2–$\overset{|}{CH}$—CH_3
  4   3   2   1

Butane-2-sulfinic acid

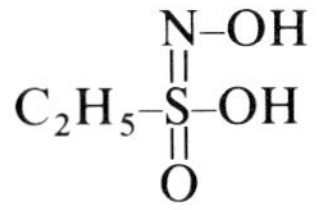

N–OH
||
C_2H_5-S–OH
||
O

Ethanesulfonohydroximic acid

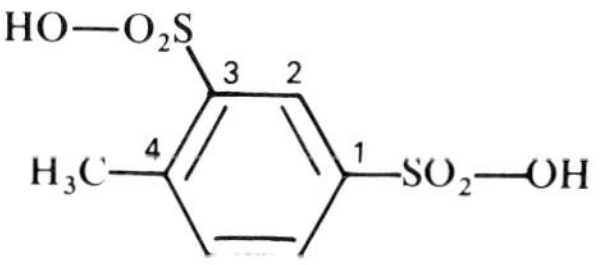

HO—O_2S

H_3C—$\overset{4}{}^{3}^{2}$—SO_2—OH

4-Methylbenzene-1,3-disulfonic acid
(**not** Toluene-2,4-disulfonic acid, see
R-9.1 Table 19(a), p. 163)

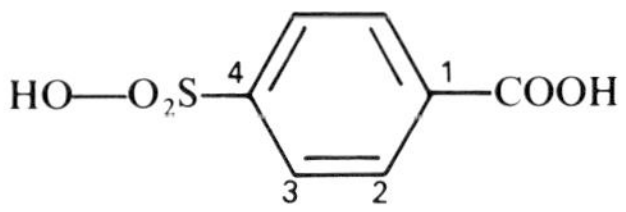

HO—O_2S—$\overset{4}{}\overset{1}{}$—COOH
        3   2

4-Sulfobenzoic acid

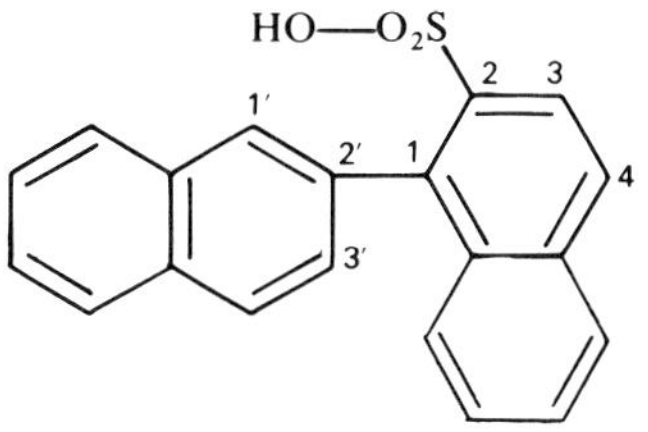

HO—O_2S

[1,2'-Binaphthalene]-2-sulfonic acid

Some trivial names are allowed (see R-9.1, Table 30, p. 178).

R-5.7.2.2   ***Selenium acids containing selenium directly linked to an organic group*** are named in the same way as the corresponding sulfur acids (see R-5.7.2.1) by replacing the stem 'sulf-' with 'selen-' in the name.

Example:
C_6H_5–SeO_2–OH
Benzeneselenonic acid

R-5.7.3   **Phosphorus and arsenic acids containing phosphorus or arsenic atoms directly linked to an organic group**

R-5.7.3.1   ***Phosphorus oxo acids and replacement modifications*** containing pentavalent phosphorus directly attached to an organic group are named substitutively as derivatives of functional parent compounds as illustrated in Table 15. Other replacement modifications are formed using the affixes given in Table 8, p. 66. Prefixes for phosphorus groups are illustrated in Table 7, p. 65.

Examples:

$(C_6H_5)_2P(O)OH$
Diphenylphosphinic acid

$(HO)_2(O)P$–CH_2–COOH
Phosphonoacetic acid

$C_2H_5P(O)(OH)_2$
Ethylphosphonic acid

Table 15 Functional parent compounds for phosphorus acids and functional replacement modifications

Parent compound	Replacement of –OH by –SH and/or =O by =S	Replacement of =O by =NH
$\overset{O}{\underset{\parallel}{\text{H}_2\text{P–OH}}}$ Phosphinic acid	$\overset{S}{\underset{\parallel}{\text{H}_2\text{P–OH}}}$ Phosphinothioic *O*-acid	$\overset{NH}{\underset{\parallel}{\text{H}_2\text{P–OH}}}$ Phosphinimidic acid
	$\overset{O}{\underset{\parallel}{\text{H}_2\text{P–SH}}}$ Phosphinothioic *S*-acid	
	$\overset{S}{\underset{\parallel}{\text{H}_2\text{P–SH}}}$ Phosphinodithioic acid	
$\overset{O}{\underset{\parallel}{\text{HP–OH}}}$ OH Phosphonic acid	$\overset{S}{\underset{\parallel}{\text{HP–OH}}}$ OH Phosphonothioic *O,O'*-acid	$\overset{NH}{\underset{\parallel}{\text{HP–SH}}}$ OH Phosphonimidothioic acid
	$\overset{O}{\underset{\parallel}{\text{HP–OH}}}$ SH Phosphonothioic *O,S*–acid	
	$\overset{O}{\underset{\parallel}{\text{HP–SH}}}$ SH Phosphonodithioic *S,S'*-acid	
	$\overset{S}{\underset{\parallel}{\text{HP–SH}}}$ SH Phosphonotrithioic acid	

R-5.7.3.2   ***Arsenic oxo acids and replacement modifications*** containing pentavalent arsenic directly attached to an organic group are named in the same way as their phosphorus analogues (see R-5.7.3.1).

Examples:

$(\text{CH}_3)_2\text{As(O)OH}$
Dimethylarsinic acid

$\text{C}_6\text{H}_5\text{As(O)(OH)}_2$
Phenylarsonic acid

$\overset{\text{N–CH}_3}{\underset{\parallel}{(\text{C}_6\text{H}_5)_2\text{As–OH}}}$

N-Methyl-*As,As*-diphenylarsinimidic acid

R-5.7.4 **Salts and esters**

R-5.7.4.1 ***Salts***
Neutral salts of organic acids are named by citing the cation(s) followed by the name of the anion as a separate word. Diverse cations are cited in alphabetical order.

Examples:

$CH_3 [CH_2]_5-COO^- K^+$
Potassium heptanoate

$CH_3-CH_2-CH_2-CH_2-CSS^- K^+$
Potassium pentanedithioate

$(CH_3-COO^-)_2 Ca^{2+}$
Calcium diacetate

$C_6H_5-SO_2^- Na^+$
Sodium benzenesulfinate

$K^+ {}^-OOC-[CH_2]_2-COO^- Na^+$
Potassium sodium succinate

$K^+ {}^-OOC-[CH_2]_5-COO^- NH_4^+$
Ammonium potassium heptanedioate

Acid salts of polybasic organic acids are named in the same way as the neutral salts, the remaining acid hydrogen atom(s) being indicated by the word 'hydrogen' (or 'dihydrogen', etc., as appropriate) inserted between the name(s) of the cation(s) and the name of the anion from which it is separated by spaces[86]. Ionic substituents such as $-COO^-$, $-SO_2O^-$, and $-PO(O^-)_2$ are described by prefix names such as 'carboxylato-', 'sulfonato-', and 'phosphonato-', respectively.

Examples:

$HOOC-[CH_2]_5-COO^- K^+$
Potassium hydrogen heptanedioate

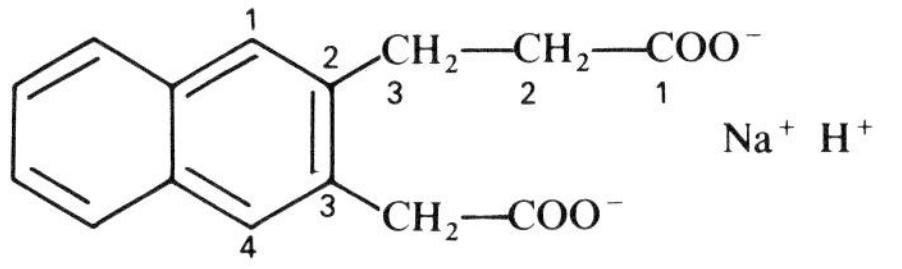

Sodium hydrogen 3-[3-(carboxylatomethyl)-2-naphthyl]propanoate

R-5.7.4.2 ***Esters.*** Fully esterified acids are named in the same way as neutral salts except that names of alkyl or aryl, etc., groups, cited in alphabetical order when more than one, replace the names of the cations.

Examples:

$CH_3-CO-O-C_2H_5$
Ethyl acetate

$PO(O-CH_3)_3$
Trimethyl phosphate

$C_2H_5-O-CO-CH_2-CO-O-CH_3$
Ethyl methyl malonate

$HPO(O-CH_3)_2$
Dimethyl phosphonate

---

[86] In the IUPAC *Nomenclature of Inorganic Chemistry*[3], it is recommended (I-8.3.3.4, pp. 109–111) that for acid salts derived from inorganic acids the word hydrogen should be joined to the following anion name, as potassium hydrogensulfate.

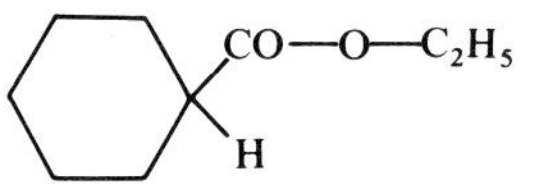

Ethyl cyclohexanecarboxylate

$CH_3-PO(O-C_2H_5)_2$
Diethyl methylphosphonate

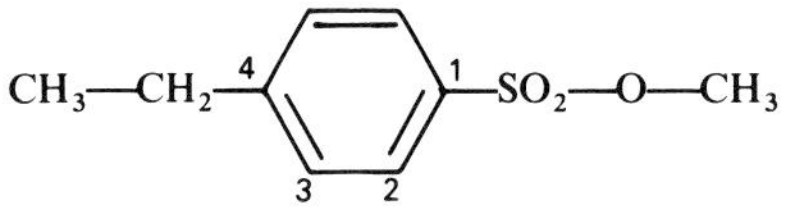

Methyl 4-ethylbenzenesulfonate

Structural specificity for esters of thio- or selenocarboxylic acids is provided by the appropriate italic element symbol, such as *S*- or *O*-, prefixed to the name of the alkyl, aryl, etc., group, as needed.

Examples:

$CH_3-[CH_2]_4-CO-S-C_2H_5$
S-Ethyl hexanethioate

$CH_3-[CH_2]_4-CSe-O-C_2H_5$
O-Ethyl hexaneselenoate

Functional class nomenclature is often used for naming esters of natural products and in index nomenclature.

Examples:

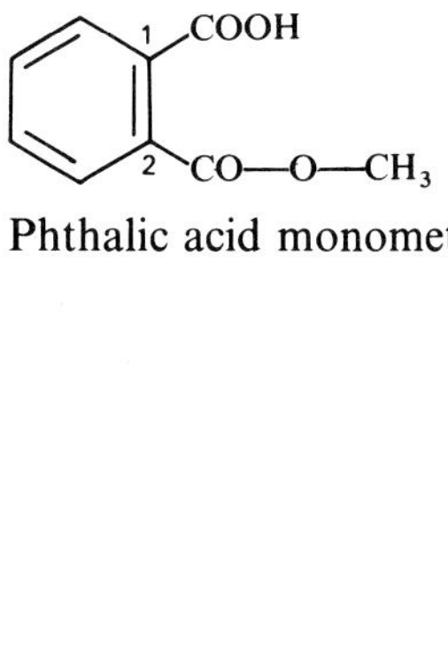

Phthalic acid monomethyl ester

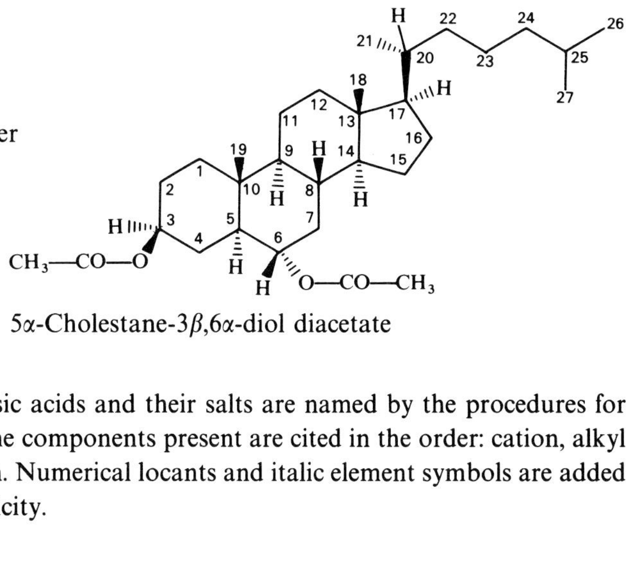

5α-Cholestane-3β,6α-diol diacetate

Partial (acid) esters of polybasic acids and their salts are named by the procedures for neutral esters and acid salts; the components present are cited in the order: cation, alkyl or aryl group, hydrogen, anion. Numerical locants and italic element symbols are added as necessary to provide specificity.

Examples:
$C_2H_5-O-CO-CH_2-CH_2-COO^- Na^+$
Sodium ethyl succinate

$C_2H_5-S-CO-CH_2-CH_2-C \begin{Bmatrix} O \\ S \end{Bmatrix}^- Li^+$
Lithium *S*-ethyl butanebis(thioate)

118

$$\begin{array}{c} COO^- \\ | \end{array}$$
$$C_2H_5{-}O{-}CO{-}CH_2{-}C(OH){-}CH_2{-}COO^-K^+H^+ \qquad\qquad HO{-}SO_2{-}O{-}C_2H_5$$
$$\phantom{C_2H_5{-}O{-}CO{-}}_5\phantom{{-}CH_2{-}}_4\phantom{{-}C(OH)}_3\phantom{{-}CH_2}_2\phantom{{-}COO}_1$$

Potassium 5-ethyl hydrogen citrate[86] Ethyl hydrogen sulfate[86]

1-Ethyl hydrogen 3-chlorophthalate[86]
(note that the numbering of the diacid is retained)

When, in an ester with the general structure R–CO–O–R′, another group is present that has priority for citation as a suffix (see Table 10, R-4.1, and Table 5, R-3.2.1.1), or when all ester groups cannot be described by the above methods, an ester group is indicated by prefixes such as 'alkoxycarbonyl-' or 'aryloxycarbonyl-' for the group –CO–OR′, or 'acyloxy–' for the group R–CO–O–.

Examples:

$$C_2H_5{-}O{-}CO{-}CH_2{-}CH_2{-}N^+(CH_3)_3 \; Br^-$$
$$\phantom{C_2H_5{-}O{-}CO{-}CH}_2\phantom{{-}CH}_1$$

[2-(Ethoxycarbonyl)ethyl]trimethylammonium bromide

$$C_6H_5{-}CO{-}O{-}CH_2{-}CH_2{-}COOH$$
$$\phantom{C_6H_5{-}CO{-}O}_3\phantom{{-}CH}_2\phantom{{-}CH}_1$$

3-(Benzoyloxy)propanoic acid

Functional class nomenclature can also be used, particularly in indexes, to describe an ester in the presence of a group with priority for citation as a principal characteristic group.

Example:

4-Hydroxybenzoic acid acetate

R-5.7.5 **Lactones, lactams, lactims, and analogues**
Compounds that may be considered as derived from a hydroxy carboxylic acid or amino carboxylic acid by loss of water intramolecularly are called generically 'lactones' or 'lactams', respectively. Tautomeric forms of lactams are called 'lactims'. In these recommendations, such compounds are preferably named as heterocycles although names that may be considered to be derived from the corresponding hydroxy or amino acid are also given.

R-5.7.5.1   *Lactones.* Intramolecular esters of hydroxy carboxylic acids are 'lactones' and are named as heterocycles or by substituting '-olactone' for the '-ic acid' ending of a trivial name of a hydroxy acid, or '-lactone' for the '-ic acid' ending of a systematic '-oic acid' name for the nonhydroxylated parent acid, and inserting a locant designating the position of the hydroxy group between the 'o' and 'lactone'[87].

Example:

Tetrahydrofuran-2-one
Butano-4-lactone
(traditionally γ-Butyrolactone)

Heterocycles in which one or more (but not all) rings of a polycyclic ring system are lactones are named by adding the suffix '-carbolactone' (denoting a cyclic –O–CO– group) to the name of the ring system left after the –O–CO– residue is replaced by two hydrogen atoms, preceded by a pair of locants indicating the points of attachment of the carbonyl group and the oxygen atom of the lactone, respectively; the locant for the carbonyl group is cited first, and, if there is a choice, is the lower locant. Multiplying prefixes and pairs of locants separated by a colon denote the presence of two or more lactone rings.

Examples:

Phenanthro[1,10-*bc*:9,8-*b'c'*]difuran-4,6-dione
Phenanthrene-1,10:9,8-dicarbolactone

8-Oxo-7-oxabicyclo[4.2.0]octane-4,5-dicarboxylic acid (R-2.4.2.1)
2-Oxohexahydro-2*H*-benzooxete-5,6-dicarboxylic acid (R-2.4.4.1)

R-5.7.5.2   *Sultones.* Intramolecular esters of hydroxy sulfonic acids are called 'sultones' and are named as heterocycles or by citing the term 'sultone' denoting the cyclic –O–SO$_2$– group

---

[87] 'Olide' names for lactones contained in previous editions of the IUPAC *Nomenclature of Organic Chemistry*[1] are not included in these recommendations.

after the name of the appropriate parent hydride preceded by a pair of locants describing the points of attachment of the sulfonyl group and oxygen atom, respectively; the locant for the sulfonyl group is cited first and, if there is a choice, is the lower locant. Multiplying prefixes and pairs of locants separated by a colon are used to indicate two or more sultone rings.

Examples:

Naphtho[1,8-*cd*][1,2]oxathiole 2,2-dioxide
Naphthalene-1,8-sultone

3-Methyl-1,2-oxathiane 2,2-dioxide
Pentane-2,5-sultone

R-5.7.5.3   **Lactams and lactims.** Nitrogen analogues of lactones having the group –CO–NH– as part of a ring or ring system are called generically 'lactams' and their tautomers, –C(OH)=N–, are 'lactims'. These compounds are named as heterocyclic compounds or in accordance with R-5.7.5.1 using '-lactam' or '-lactim', respectively, in place of '-lactone'.

Names such as 'propiolactam' and 'butyrolactim' are not included in these recommendations.

Examples:

Tetrahydropyrrol-2-one
Pyrrolidin-2-one (2-Pyrrolidone[88])
Butano-4-lactam

3,4-Dihydro-2*H*-pyrrol-5-ol
Butano-4-lactim

R-5.7.5.4   **Sultams.** Nitrogen analogues of sultones having the group –SO₂–NH– as a part of a ring are named as heterocycles or in accordance with R-5.7.5.2 using 'sultam' in place of 'sultone'. The locant for the point of attachment of the sulfonyl group is cited first and, where there is a choice, has preference over the imino group for lower locant.

Examples:

2*H*-Naphtho[1,8-*cd*]isothiazole 1,1-dioxide
Naphthalene-1,8-sultam

1,2-Thiazinane 1,1-dioxide
Butane-1,4-sultam

[88] The '-idinone' suffix has often been abbreviated to '-idone', for example, piperidone, isoxazolidone, and imidazolidone (see also R-9.1, Table 27(a), p. 174).

R-5.7.6   **Acid halides**

Acid halides in which hydroxy groups of all acid groups expressed as the principal characteristic group (carboxylic, sulfonic, sulfinic, selenonic, etc., acids) have been replaced by halogen atoms are named by citing the name of the acyl group (see R-5.7.1.1) followed by the name(s) of the specific halide(s) as separate words, in alphabetical order, each preceded by a multiplicative prefix, as needed.

Examples:

CH_3–CO–Cl
Acetyl chloride

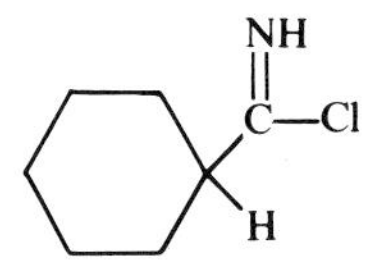

Cyclohexanecarboximidoyl chloride

CH_3–[CH_2]$_4$–CO–Br
Hexanoyl bromide

C_6H_5–SO–Cl
Benzenesulfinyl chloride

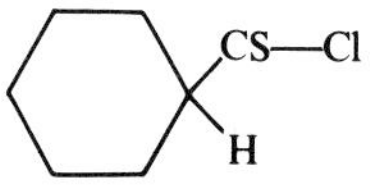

Cyclohexanecarbothioyl chloride

CH_3–CH_2–SO_2–Br
Ethanesulfonyl bromide

Br–CO–CH_2–CO–Cl
Malonyl bromide chloride

C_6H_5–SeO–Cl
Benzeneseleninyl chloride

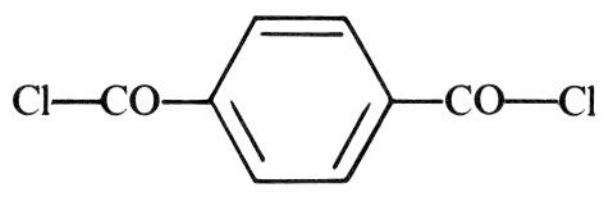

Terephthaloyl dichloride

When another group is present that has priority for citation as principal group or when attached to a substituting group, an acyl halide group is expressed by a prefix such as 'fluorocarbonyl-', 'chlorocarbonyl-', 'bromocarbonyl-' or 'iodocarbonyl-'[89].

Examples:

Cl–CO–CH_2–COOH
(Chlorocarbonyl)acetic acid

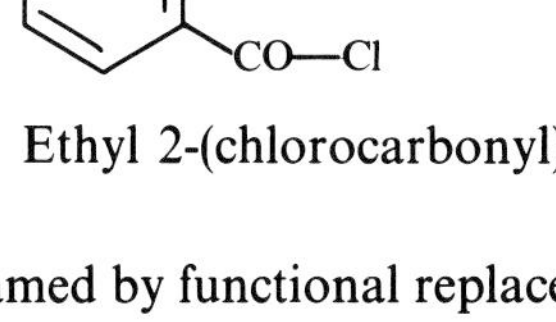

Ethyl 2-(chlorocarbonyl)benzoate

Mononuclear oxo acid halides and analogues named by functional replacement nomenclature (see R-3.4), which may be named as acyl halides as described above, are often

---

[89] Prefixes such as 'fluoroformyl-', 'chloroformyl-' 'bromoformyl-', and 'iodoformyl' were used in the 1979 edition of the IUPAC *Nomenclature of Organic Chemistry*[1].

named by replacing the term 'acid' of the acid name by the appropriate halide class name(s).

Examples:

$C_6H_5-P(O)Cl_2$
Phenylphosphonic dichloride
Phenylphosphonoyl dichloride

$(C_2H_5)_2P(S)-Cl$
Diethylphosphinothioic chloride
Diethylphosphinothioyl chloride

$\overset{\displaystyle N-C_6H_5}{\underset{(C_2H_5)_2\overset{\|}{P}-Cl}{}}$

P,P-Diethyl-*N*-phenylphosphinimidic chloride
P,P-Diethyl-*N*-phenylphosphinimidoyl chloride

R-5.7.7 **Anhydrides and their analogues**
Anhydrides are compounds formally derived by the loss of water from two acid functions[90].

R-5.7.7.1 *Symmetrical anhydrides.* Symmetrical anhydrides of monobasic acids, substituted or unsubstituted, are named by replacing the term 'acid' of the acid name by the class name 'anhydride'[91].

Examples:

$(CH_3-CO)_2O$
Acetic anhydride

$(CH_3-CH_2-CH_2-CH_2-CH_2-CO)_2O$
Hexanoic anhydride

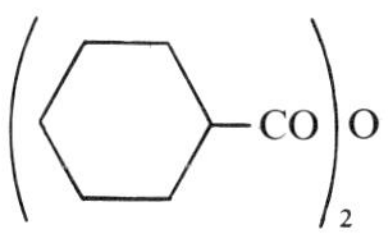

Cyclohexanecarboxylic anhydride

$(ClCH_2-CO)_2O$
Chloroacetic anhydride

$(C_6H_5-CS)_2O$
(Thiobenzoic) anhydride[92]

$(CH_3-CH_2-CS)_2O$
(Thiopropionic) anhydride

$(C_6H_5-SO_2)_2O$
Benzenesulfonic anhydride

$(ClCH_2-CH_2-SO)_2O$
2-Chloroethanesulfinic anhydride

Cyclic anhydrides formed from two acid groups attached to the same parent hydride structure may be named in the same manner as symmetrical anhydrides, or as heterocyclic compounds.

[90] Accordingly, an anhydrosulfide is a compound formally derived by the loss of hydrogen sulfide from two thio acid groups.
[91] In previous recommendations[1], the multiplicative prefix 'bis-', although optional, was used in the names of symmetrically substituted anhydrides.
[92] The use of the parentheses to enclose the name of the thio acid residue removes the possibility of ambiguity in this example and the following.

Examples:

Succinic anhydride
Tetrahydrofuran-2,5-dione

Phthalic anhydride
1,3-Dihydro-2-benzofuran-1,3-dione[93]
1,3-Dihydrobenzo[*c*]furan-1,3-dione
1,3-Dihydroisobenzofuran-1,3-dione

Cyclohexane-1,2,3,4-tetracarboxylic acid 3,4-anhydride
1,3-Dioxooctahydro-2-benzofuran-4,5-dicarboxylic acid[93]
1,3-Dioxooctahydrobenzo[*c*]furan-4,5-dicarboxylic acid
1,3-Dioxooctahydroisobenzofuran-4,5-dicarboxylic acid

R-5.7.7.2   ***Unsymmetrical (mixed) anhydrides.*** Anhydrides derived from different monobasic acids
are named by citing the first parts of the names of the two acids (i.e., the parts preceding
the term acid) in alphabetical order, followed by the class name 'anhydride' as a separate
word.

Examples:

$CH_3-CO-O-CO-CH_2-CH_3$
Acetic propionic anhydride

$CH_3-CS-O-CO-C_6H_5$
Benzoic thioacetic anhydride

$CH_3-CO-O-CO-CH_2Cl$
Acetic chloroacetic anhydride

$C_2H_5-SO_2-O-SO-C_6H_5$
Benzenesulfinic ethanesulfonic anhydride

$ClCH_2-CO-O-SO_2-$ (1) (4)$-NO_2$ (2)(3)

Chloroacetic 4-nitrobenzenesulfonic anhydride

R-5.7.7.3   ***Chalcogen analogues of anhydrides*** having the general structure –CO–S–CO–,
–CO–S–CS–, or –CS–S–CS– are named in the same way as their oxygen analogues,
using the class name 'thioanhydride'. The sulfur linkage between the acyl groups is given
by the replacement prefix 'thio-'; other sulfur atoms are indicated by 'thio-' prefixes cited
with the appropriate acyl group.

     Alternatively, such thioanhydrides may be named on the basis of the parent hydride
name 'sulfane', the corresponding class name being 'diacylsulfanes'.

---

[93] See Rule B-3.5, 1979 edition, IUPAC *Nomenclature of Organic Chemistry*[1], p. 67.

Examples:

(C₆H₅–CO)₂S → $(C_6H_5-CO)_2S$
Benzoic thioanhydride
Dibenzoylsulfane

$C_2H_5-SO_2-S-CS-C_6H_5$
Ethanesulfonic thiobenzoic thioanhydride
Ethanesulfonyl(thiobenzoyl)sulfane

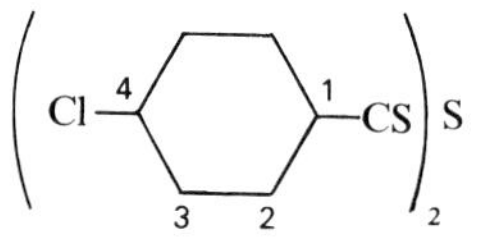

4-Chlorocyclohexane-1-carbothioic thioanhydride
Bis(4-chlorocyclohexane-1-carbothioyl)sulfane

The various unsymmetrical thioanhydrides derived from acetic propionic anhydride are named as follows:

$CH_3-CO-O-CO-CH_2-CH_3$
Acetic propionic anhydride

$CH_3-CO-S-CO-CH_2-CH_3$
Acetic propionic thioanhydride
Acetyl(propionyl)sulfane

$CH_3-CO-O-CS-CH_2-CH_3$
Acetic thiopropionic anhydride

$CH_3-CS-O-CO-CH_2-CH_3$
Propionic thioacetic anhydride

$CH_3-CO-S-CS-CH_2-CH_3$
Acetic thiopropionic thioanhydride
Acetyl(thiopropionyl)sulfane

$CH_3-CS-S-CO-CH_2-CH_3$
Propionic thioacetic thioanhydride
Propionyl(thioacetyl)sulfane

$CH_3-CS-O-CS-CH_2-CH_3$
Thioacetic thiopropionic anhydride

$CH_3-CS-S-CS-CH_2-CH_3$
Thioacetic thiopropionic thioanhydride
Thioacetyl(thiopropionyl)sulfane

R-5.7.8   **Amides, imides, and hydrazides**

R-5.7.8.1   *Monoacyl derivatives of ammonia (primary amides[94])* are named by replacing the suffixes '-oic acid', '-ic acid', or '-carboxylic acid' of the name of the acid corresponding to the acyl group by '-amide' or '-carboxamide'. The substituent prefix corresponding to –CONH₂ is 'carbamoyl-'.

Examples:

$CH_3-CO-NH_2$
Acetamide

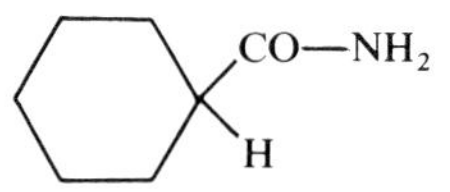

Cyclohexanecarboxamide

[94] Although unsubstituted monoacyl derivatives of ammonia can be called 'primary amides', classification of mono-, di-, and triacyl derivatives of ammonia as primary, secondary, and tertiary amides in analogy with amines cannot be encouraged because of the common usage of the term 'tertiary amide' to describe a disubstituted primary amide with the general structure R–CO–NR'R''.

CH$_3$–CH$_2$–CH$_2$–CH$_2$–CH$_2$–CO–NH$_2$
Hexanamide

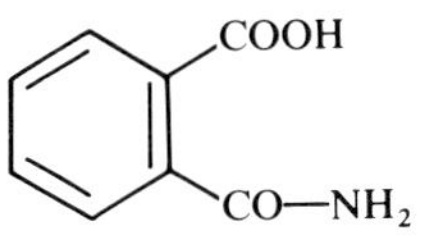

2-Carbamoylbenzoic acid
Phthalamic acid (see R-5.7.1.2.2)

C$_6$H$_5$–SO$_2$–NH$_2$
Benzenesulfonamide

Some trivial names are retained (see R-9.1, Table 28(a), p. 175).

Substituted primary amides with general structures such as R–CO–NHR′, R–CO–NR′R″, and the corresponding sulfonamides, are named by citing the substituents R′ and R″ as prefixes (but see below for *N*-phenyl derivatives).

Examples:

C$_6$H$_5$–CO–NH–CH$_3$
N-Methylbenzamide

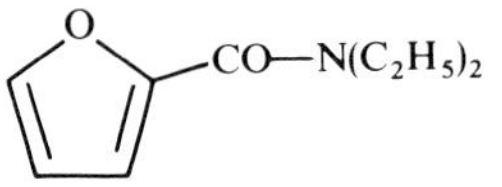

N,N-Diethyl-2-furamide
N,N-Diethylfuran-2-carboxamide

N-Phenyl derivatives of primary amides are called 'anilides' and may be named using the suffix '-anilide' in place of the suffix '-amide'. Locants for substituents in the *N*-phenyl ring are primed.

Examples:

CH$_3$–CO–NH–C$_6$H$_5$
Acetanilide
N-Phenylacetamide

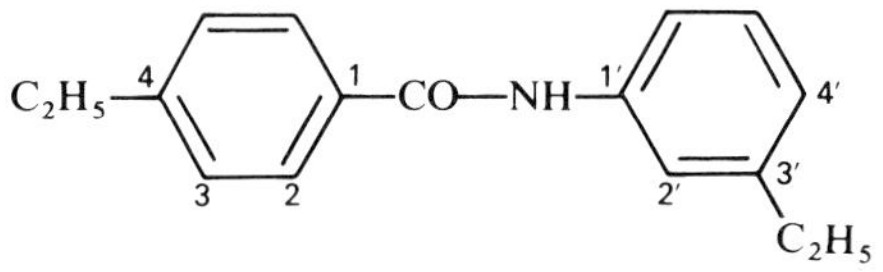

3′,4-Diethylbenzanilide

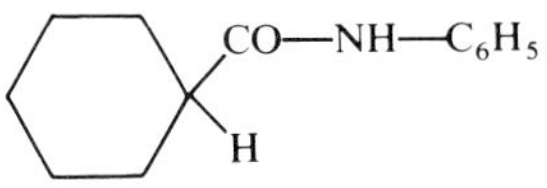

Cyclohexanecarboxanilide

Alternatively, an *N*-acyl group may be named as a *N*-substituent of an amine. This method is used mainly for *N*-derivatives of nitrogenous heterocycles.

Example:

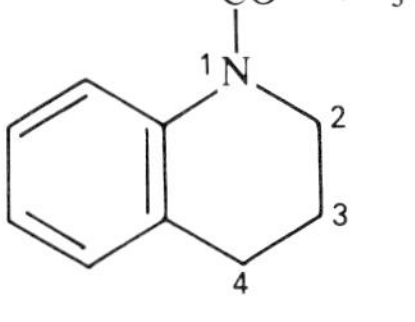

1-Acetyl-1,2,3,4-tetrahydroquinoline

When a group having preference for citation as a principal characteristic group is present, the group R–CO–NH– of an *N*-substituted amide R–CO–NH–R′ may be

126

treated as a substituent of the compound HR′. The group R–CO–NH– may be
expressed as a substituent by changing the '-amide' or '-carboxamide' suffix of the amide
name to 'amido-' or 'carboxamido-', respectively, or alternatively, by the appropriate
'acylamino-' prefix.

Examples:

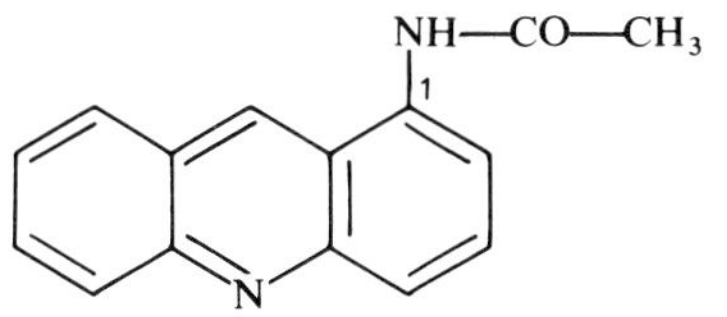

1-Acetamidoacridine
1-(Acetylamino)acridine

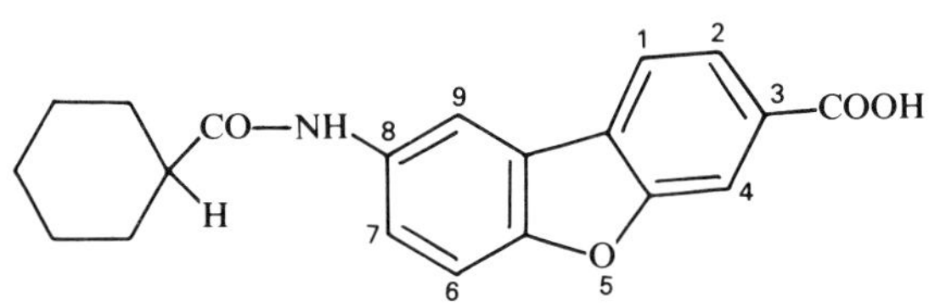

8-(Cyclohexanecarboxamido)dibenzofuran-3-carboxylic acid
8-[(Cyclohexanecarbonyl)amino]dibenzofuran-3-carboxylic acid

R-5.7.8.2   ***Symmetrical diacyl and triacyl derivatives of ammonia,*** i.e., $(R–CO)_2NH$ and $(R–CO)_3N$,
respectively, are named as di- and triacyl derivatives of the parent hydride azane, or as di-
and triacylamines.

Examples:

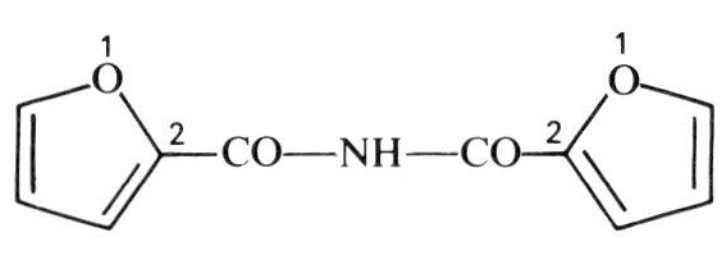

Di-2-furoylazane   Tris(cyclohexanecarbonyl)azane
Di-2-furoylamine   Tris(cyclohexanecarbonyl)amine

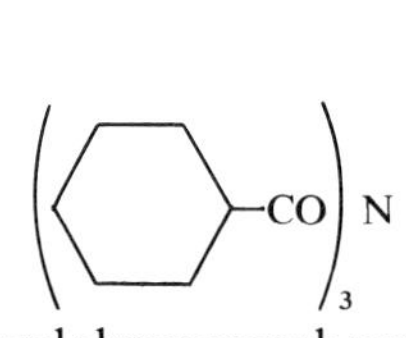

The trivial names diacetamide and triacetamide for the compounds $(CH_3–CO)_2NH$ and
$(CH_3–CO)_3N$, respectively, and dibenzamide and tribenzamide for $(C_6H_5–CO)_2NH$ and
$(C_6H_5–CO)_3N$, respectively, have been used but are not included in these recommendations for use in naming carbon-substituted derivatives.

Unsymmetrical di- and triacyl derivatives of ammonia, such as R–CO–NH–CO–R′,
$R–N(CO–R′)_2$, and R–CO–N(CO–R′)(CO–R″), may be named (a) on the basis of the
parent hydride azane; (b) as a diacylamine, a diacylalkyl- (or aryl, etc.) amine, or a
triacylamine; or (c) as alkyl (or aryl, etc.) or acyl derivatives of a primary amide (see
R-5.7.8.1), diacetamide, or dibenzamide.

Examples:

C₆H₅–CO–NH–CO–CH₃
(a) Acetyl(benzoyl)azane
(b) Acetyl(benzoyl)amine
(c) *N*-Acetylbenzamide

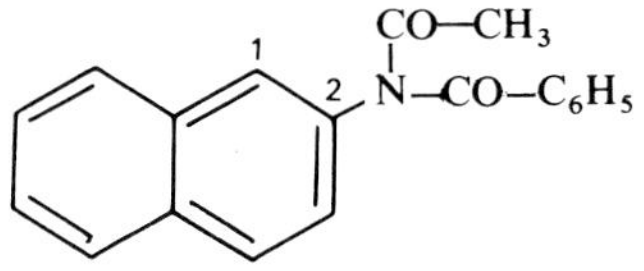

(a) Acetyl(benzoyl)-2-naphthylazane
(b) Acetyl(benzoyl)-2-naphthylamine
(c) *N*-Acetyl-*N*-(2-naphthyl)benzamide

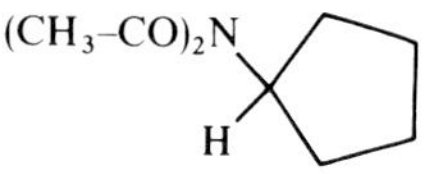

(CH₃–CO)₂N

(a) Diacetyl(cyclopentyl)azane
(b) Diacetyl(cyclopentyl)amine
(c) *N*-Cyclopentyldiacetamide

$$ \underset{\underset{\text{ClCH}_2\text{–CH}_2\text{–CO–N–CO–CH}_3}{|}}{\text{CO–C}_6\text{H}_5} $$

(a) Acetyl(benzoyl)(3-chloropropanoyl)azane
(b) Acetyl(benzoyl)(3-chloropropanoyl)amine
(c) *N*-Acetyl-*N*-(3-chloropropanoyl)benzamide

R-5.7.8.3   ***Imides*** are compounds containing the structural grouping –CO–NH–CO– and may be considered as nitrogen analogues of anhydrides or as diacyl derivatives of ammonia. Acyclic imides are named according to R-5.7.8.2. Cyclic imides are named by replacing the suffixes '-dioic acid', '-ic acid', or '-dicarboxylic acid' of the corresponding dibasic acid by '-imide' or '-dicarboximide'. They can also be named as heterocycles.

Examples:

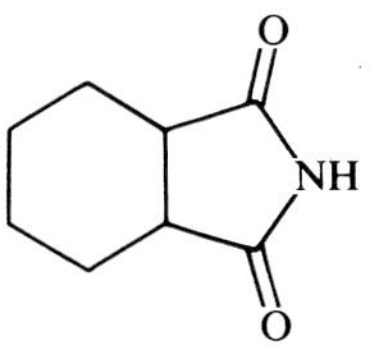

Cyclohexane-1,2-dicarboximide
Octahydroisoindole-1,3-dione

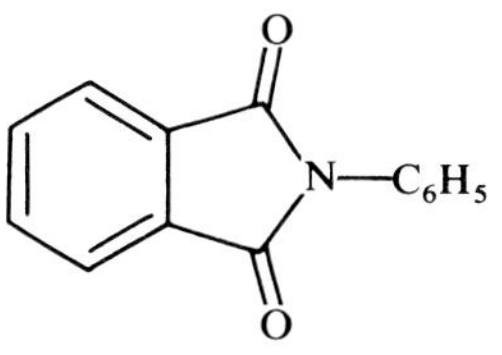

N-Phenylphthalimide
2-Phenyl-2,3-dihydro-1*H*-isoindole-1,3-dione

R-5.7.8.4   ***Hydrazides***. Monoacyl derivatives of the parent hydride diazane (hydrazine), H₂N–NH₂, have the generic name 'hydrazides' and are named by replacing the '-ic acid' or '-oic acid' suffix of the name of the acid by '-ohydrazide' or the suffix '-carboxylic acid' by '-carbohydrazide'.

Examples:

C₆H₅–CO–NH–NH₂
Benzohydrazide

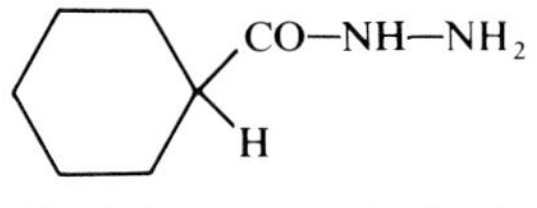

Cyclohexanecarbohydrazide

CH₃–CH₂–CH₂–CH₂–CO–NH–NH₂
Pentanohydrazide

Alkyl, aryl, cycloalkyl, etc., substituents on the nitrogen atoms of hydrazides are described by the appropriate prefix names using the locants *N*- (or 1'-) for the imino nitrogen atom and *N'*- (or 2'-) for the amino nitrogen atom. Hydrazide derivatives may also be named as derivatives of the parent hydride diazane (hydrazine).

Examples:

CH₃
|
CH₃–CO–N–NH₂

N-Methylacetohydrazide
1'-Methylacetohydrazide
1-Acetyl-1-methyldiazane (hydrazine)

CH₃
|
CH₃–CH₂–CO–N–N(CH₃)₂

N,N',N'-Trimethylpropionohydrazide
1',2',2'-Trimethylpropionohydrazide
1,1,2-Trimethyl-2-propionyldiazane (hydrazine)

Diacyl, triacyl and tetraacyl derivatives of hydrazides are named on the basis of the parent hydride diazane (hydrazine).

Examples:

C₆H₅–CO–NH–NH–CO–C₆H₅
          2   1

1,2-Dibenzoyldiazane (hydrazine)

H₃C   CH₂–CH₃
 | |
CH₃–CH₂–CO–N–N–CO–CH₃
             2   1

1-Acetyl-1-ethyl-2-methyl-2-propionyldiazane (hydrazine)

When a group having priority for citation as a principal characteristic group is present, a hydrazide group is expressed by the prefix 'acyldiazanyl-' (acylhydrazino-). Locants for positions on the hydrazino- group are *N*- (or 1-) for the nitrogen atom adjacent to the free valence position and *N'*- (or 2-) for the other nitrogen atom.

Example:

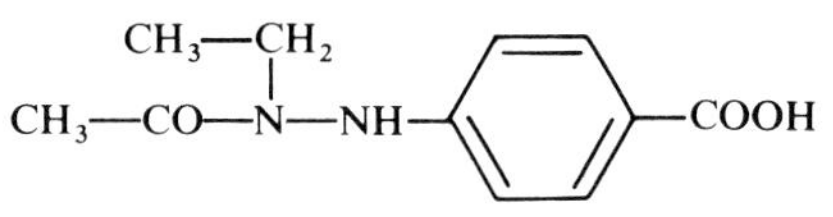

CH₃—CH₂
        |
CH₃—CO—N—NH—⟨benzene ring⟩—COOH

4-(2-Acetyl-2-ethyldiazanyl)benzoic acid
4-(*N'*-Acetyl-*N'*-ethylhydrazino)benzoic acid
4-(2-Acetyl-2-ethylhydrazino)benzoic acid

R-5.7.9   **Nitriles, isocyanides and related compounds**[95]

R-5.7.9.1   *Nitriles.* Compounds with the general structure R–C≡N are called 'nitriles' or 'cyanides' generically and may be named substitutively in a manner closely related to that for acids and other related compounds. Acyclic mono- and dinitriles in which –C≡N may be considered to have replaced the –COOH group(s) of an acid named by an '-oic acid' or '-dioic acid' suffix are named by adding the suffix '-nitrile' or '-dinitrile' to the name of the hydrocarbon from which the acid name was derived.

[95] Cyanide, isocyanide, cyanate and related groups (see Table 16) are often considered as 'pseudohalides' and thus may be treated in the same way as their halogen analogues (see R-5.3.1).

Examples:

CH$_3$–CH$_2$–CH$_2$–CH$_2$–C≡N
Pentanenitrile

N≡C–CH$_2$–CH$_2$–CH$_2$–CH$_2$–C≡N
Hexanedinitrile

Nitriles in which –C≡N may be considered to have replaced –COOH of an acid with a retained trivial name (see R-9.1, Table 28(a), p. 175) are named by changing the '-ic acid', or '-oic acid' ending of the name of the acid to '-onitrile'.

Examples:

CH$_3$–C≡N
Acetonitrile

N≡C–CH$_2$–CH$_2$–C≡N
Succinonitrile

C$_6$H$_5$–C≡N
Benzonitrile

Nitriles in which –C≡N may be considered to have replaced the –COOH group(s) of an acid named by a '-carboxylic acid' suffix are named by replacing that suffix with '-carbonitrile'.

Examples:

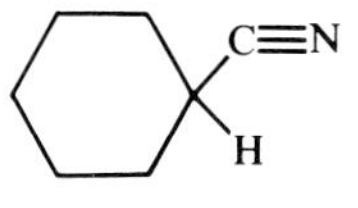

Cyclohexanecarbonitrile

N≡C–CH$_2$–CH$_2$–CH$_2$–ĊH–CH$_2$–CH$_2$–C≡N (with C≡N on position 3)
6   5   4   3   2   1
Hexane-1,3,6-tricarbonitrile

Functional class names for nitriles of the general structure R–C≡N are formed by citing the prefix name for the group R followed by the class name 'cyanide' as a separate word. Compounds with a general structure such as R–CO–C≡N and R–SO$_2$–C≡N may be named similarly.

Examples:

CH$_3$–C≡N
Methyl cyanide

C$_6$H$_5$–SO$_2$–C≡N
Benzenesulfonyl cyanide

C$_6$H$_5$–CO–C≡N
Benzoyl cyanide

When a group is present that has priority for citation as the principal characteristic group or when all –C≡N groups cannot be expressed as the principal characteristic group, –C≡N groups are described by the prefix 'cyano-'.

Examples:

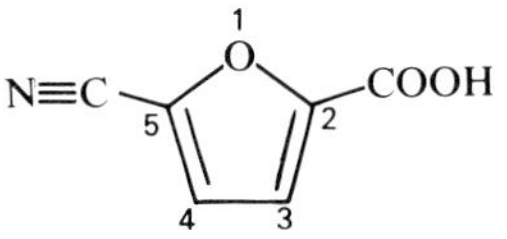

N≡C—[O ring structure]—COOH

5-Cyano-2-furoic acid
5-Cyanofuran-2-carboxylic acid

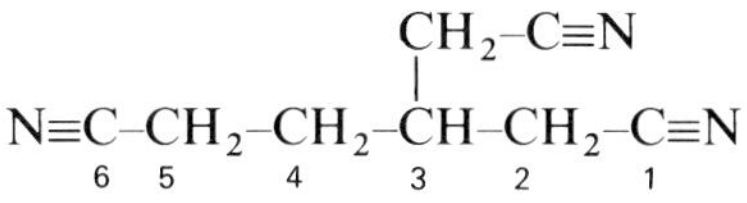

CH₂–C≡N
|
N≡C–CH₂–CH₂–CH–CH₂–C≡N
6   5   4   3   2   1

3-(Cyanomethyl)hexanedinitrile

R-5.7.9.2   *Cyanide-related compounds.* Compounds containing a group X listed in the first column of Table 16 are named by methods analogous to those described for halides (see R-5.3.1); the functional class names given in the second column of the Table are used in place of 'halide' or the prefixes given in the third column of the Table in place of 'halo-'.

Table 16 Cyanide and related groups in order of decreasing priority for citation as functional class name

Group X in RX	Functional class ending and generic class name	Prefix
–CN	cyanide	cyano-
–NC	isocyanide[a]	isocyano-
–OCN	cyanate	cyanato-
–NCO	isocyanate	isocyanato-
–ONC	fulminate	fulminato-
–SCN	thiocyanate	thiocyanato-
–NCS	isothiocyanate	isothiocyanato-
–SeCN	selenocyanate	selenocyanato-
–NCSe	isoselenocyanate	isoselenocyanato-

[a] Not isonitrile or carbylamine.

Examples:
C₆H₅–NC
Phenyl isocyanide

Cyclohexyl isocyanate

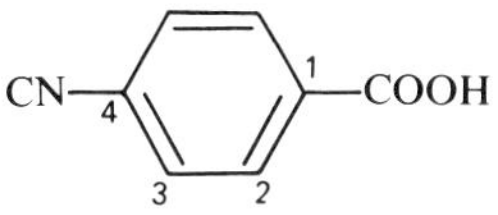

CN—[benzene ring]—COOH

4-Isocyanobenzoic acid

R-5.7.9.3   *Nitrile oxides.* Compounds with the general structure R–C≡NO have the generic name 'nitrile oxides'. In specific cases, the class name 'oxide' is added as a separate word after a nitrile name, but not a cyanide name.

Example:
C₆H₅–C≡NO
Benzonitrile oxide

R-5.8 **RADICALS AND IONS**
(For a more detailed treatment of radicals, ions, and related species, see the reference cited in footnote 45)

R-5.8.1 **Radicals**

R-5.8.1.1 *Monovalent radicals.* A radical formally derived by the removal of one hydrogen atom from any of the mononuclear parent hydrides of the carbon family of elements, or from a terminal atom of a saturated unbranched acyclic hydrocarbon, or from any position of a monocyclic hydrocarbon ring is named by replacing the '-ane' ending of the systematic name of the parent hydride by '-yl'.

Examples:

CH_3
Methyl

$CH_3–CH_2–CH_2$
Propyl

GeH_3
Germyl

Cyclobutyl

A radical formally derived by the removal of one hydrogen atom from any position of any other parent hydride, is named by adding the suffix '-yl' to the name of the parent hydride, eliding the final 'e' of the name of the parent hydride, if present.

Examples:

SH
Sulfanyl

NH_2
Azanyl
Aminyl
(traditional name: amino)

$SiH_3–\overset{\bullet}{S}iH–SiH_3$
$\quad$ 3 $\quad$ 2 $\quad$ 1

Trisilan-2-yl

Bicyclo[2.2.1]heptan-2-yl

Spiro[4.5]decan-8-yl

Cyclopenta-2,4-dien-1-yl

Note: As exceptions, the names of the radicals HO• and HOO• are 'hydroxyl' and 'hydroperoxyl', respectively.

R-5.8.1.2 *Divalent and trivalent radicals.* Divalent radical centres formally derived by removal of two hydrogen atoms from the mononuclear parent hydrides CH_4, NH_3, and SiH_4 may be named respectively 'methylene' or 'carbene'; 'azanylidene', 'nitrene', or 'aminylene'; and 'silylene'. Derivatives of these parent hydride radicals are named substitutively.

Note: The use of these names does not imply a specific electronic configuration. If needed, such a distinction should be made by using a separate word or descriptive phrase, such as singlet and triplet.

132

Examples:

$(C_6H_5)_2\overset{\bullet}{\underset{\bullet}{C}}$ and/or $(C_6H_5)_2C\colon\!\!\bullet\!\!\bullet$

Diphenylmethylene (preferred)
Diphenylcarbene

$CH_3-\overset{\bullet}{\underset{\bullet}{N}}$ and/or $CH_3-N\colon\!\!\bullet\!\!\bullet$

Methylazanylidene
Methylnitrene
Methylaminylene

$C_6H_5-CH_2-\overset{\bullet}{Si}H$
and/or
$C_6H_5-CH_2-CH_2-\overset{\bullet\bullet}{Si}H$

Benzylsilylene

$CH_3-CO-\overset{\bullet}{\underset{\bullet}{N}}$ and/or $CH_3-CO-N\colon\!\!\bullet\!\!\bullet$

Acetylazanylidene
Acetylnitrene
Acetylaminylene

Other bivalent radical centres and trivalent radical centres formally derived by removal of two or three hydrogen atoms from the same skeletal atom of a parent hydride may be described by adding a suffix '-ylidene' or '-ylidyne' following the procedure prescribed above for the suffix '-yl' (see R-5.8.1.1).

Examples:

$CH_3-\overset{\bullet}{\underset{\bullet}{C}}H$ and/or $CH_3-CH\colon\!\!\bullet\!\!\bullet$
Ethylidene

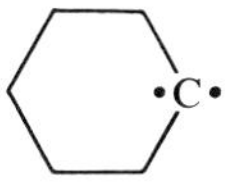

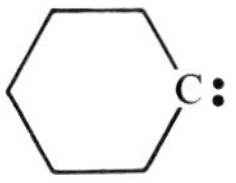

Cyclohexylidene

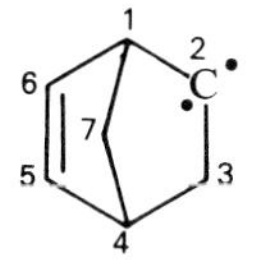

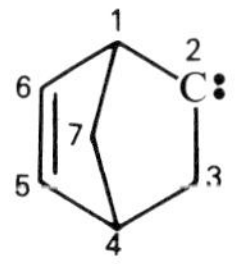

Bicyclo[2.2.1]hept-5-en-2-ylidene

Polyradicals containing two or more radical centres, formally derived by the removal of two or more hydrogen atoms from each of two or more different skeletal atoms of a parent hydride, are named by adding to the name of the parent hydride combinations of the suffixes '-yl' for a monovalent radical centre, '-ylidene' for a divalent radical centre, and '-ylidyne' for a trivalent radical centre, together with the appropriate numerical prefixes indicating the number of each kind of radical centre. The final 'e' of the name of the parent hydride, if present, is elided when followed by the letter 'y'.

Examples:

$\bullet NH-NH\bullet$
‌ 2 ‌ 1

Diazane-1,2-diyl
Hydrazine-1,2-diyl

$\bullet CH_2-\overset{\bullet}{C}H-CH_2\bullet$
‌ 3 ‌ ‌ 2 ‌ ‌ 1

Propane-1,2,3-triyl

Cyclohexan-1-yl-3-ylidene

$$CH_3-\overset{\bullet}{\underset{\bullet}{C}}-CH_2-\overset{\bullet}{\underset{\bullet}{C}}-CH_3 \text{ and/or } CH_3-\overset{\bullet}{\underset{\bullet}{C}}-CH_2-\overset{\bullet\bullet}{C}-CH_3 \text{ and/or } CH_3-\overset{\bullet\bullet}{C}-CH_2-\overset{\bullet\bullet}{C}-CH_3$$

Pentane-2,4-diylidene

$$\underset{5}{\overset{\bullet}{C}}-\underset{4}{CH_2}-\underset{3}{\overset{\bullet}{C}H}-\underset{2}{CH_2}-\underset{1}{\overset{\bullet}{C}H} \text{ and/or } \underset{5}{\overset{\bullet}{:}C}-\underset{4}{CH_2}-\underset{3}{\overset{\bullet}{C}H}-\underset{2}{CH_2}-\underset{1}{\overset{\bullet}{C}H} \text{ and/or }$$

$$\underset{5}{\overset{\bullet}{\underset{\bullet}{C}}}-\underset{4}{CH_2}-\underset{3}{\overset{\bullet}{C}H}-\underset{2}{CH_2}-\underset{1}{\overset{\bullet\bullet}{C}H} \text{ and/or } \underset{5}{\overset{\bullet}{:}C}-\underset{4}{CH_2}-\underset{3}{\overset{\bullet}{C}H}-\underset{2}{CH_2}-\underset{1}{\overset{\bullet\bullet}{C}H}$$

Pentan-3-yl-1-yliden-5-ylidyne

The presence of a radical centre in a substituent to be cited as a prefix is expressed by the prefix 'ylo-' denoting the removal of a hydrogen atom.

Example:

3-(2-Yloethyl)cyclohexyl

R-5.8.1.3   ***Radical centres on characteristic groups.*** Acyl radicals, i.e., radicals with at least one chalcogen or nitrogen atom attached to the radical centre by a (formal) double bond, which may be considered to be formally derived by the loss of the hydroxy group from acid characteristic groups, are named by replacing the '-ic acid' or '-carboxylic acid' ending of the name with '-yl' (occasionally '-oyl') or '-carbonyl'. Alternatively, acyl radicals may be named on the basis of an appropriate parent hydride using prefixes such as 'oxo-', 'thioxo-', 'imino-', etc.

Examples:

$CH_3-[CH_2]_4-\overset{\bullet}{C}S$
Hexanethioyl
1-Thioxohexyl

Benzene-1,4-disulfinyl
1,4-Phenylenebis(oxo-λ^4-sulfanyl)

Cyclohexanecarbonyl
Cyclohexyloxomethyl

$(CH_3)_2\overset{\bullet}{P}O$
Dimethylphosphinoyl
Dimethyloxo-λ^5-phosphanyl

Terephthaloyl
1,4-Phenylenebis(oxomethyl)

134

A radical derived formally by the removal of one or two hydrogen atoms from an amine, imine, or amide characteristic group may be named by adding a suffix '-aminyl', '-iminyl', or '-amidyl', to the name of the parent hydride for monovalent radicals and as a substituted nitrene for bivalent radicals. Alternatively, such radicals may be named substitutively on the basis of the parent hydride 'azane'.

Examples:

CH₃ ṄH
Methanaminyl
Methylazanyl
(traditionally: methylamino)

HCO–ṄH
Formamidyl
Formylazanyl

C₆H₅–Ṅ and/or C₆H₅–N:

Phenylazanylidene
Phenylnitrene
Phenylaminylene

Polyradicals with radical centres identically derived located on two or more amine, imine, or amide characteristic groups are named by applying the principles for nomenclature of assemblies of identical units[96] using the multiplicative prefixes 'bis-', 'tris-', etc.

Examples:

HṄ–CH₂–CH₂–ṄH
Ethylenebis(aminyl)
Ethylenebis(azanyl)

•N=C=N•
Methanediylidenebis(aminyl)
Methanediylidenebis(azanyl)

Ṅ–CO–[CH₂]₄–CO–Ṅ and/or Ṅ–CO–[CH₂]₄–CO–N: and/or :N–CO–[CH₂]₄–CO–N:

Hexanedioylbis(azanylidene)
Hexanedioylbis(nitrene)
Hexanedioylbis(aminylene)

A radical derived formally by the removal of the hydrogen atom from the hydroxy group of an acid or hydroxy characteristic group is named on the basis of a composite parent radical name formed by combining the appropriate prefix derived from the parent hydride or acid residue attached to the chalcogen atom with the term 'oxyl', 'peroxyl', etc.

Examples:

CH₃–O•
Methoxyl

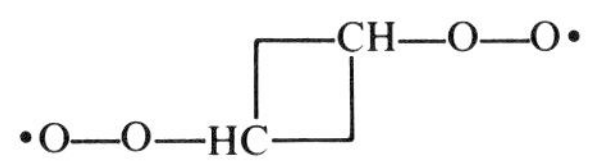

Cyclobutane-1,3-diyldiperoxyl

ClCH₂–CO–O•
Chloroacetoxyl

---

[96] See Rule C-72, p. 130, in the 1979 edition of the IUPAC *Nomenclature of Organic Chemistry*[1].

Chalcogen analogues are named on the basis of parent radicals, such as 'sulfanyl' and 'selanyl'.

Examples:

C$_6$H$_5$–S•   C$_6$H$_5$–CO–Se•
Phenylsulfanyl   Benzoylselanyl

R-5.8.2   **Cations**

A parent cation derived formally by the addition of one hydron[71] to a mononuclear hydride (see R-2.1) of the nitrogen, chalcogen, and halogen families having a standard bonding number (see R-1.1) is named by adding the suffix '-onium' to a root for the element (see Table 17). Substituents are described in the usual way.

Table 17 Mononuclear parent onium ions

Parent cation	Name	Parent cation	Name	Parent cation	Name
NH$_4^+$	ammonium[a]	OH$_3^+$	oxonium	FH$_2^+$	fluoronium
PH$_4^+$	phosphonium	SH$_3^+$	sulfonium	ClH$_2^+$	chloronium
AsH$_4^+$	arsonium	SeH$_3^+$	selenonium	BrH$_2^+$	bromonium
SbH$_4^+$	stibonium	TeH$_3^+$	telluronium	IH$_2^+$	iodonium
BiH$_4^+$	bismuthonium				

[a] The name 'nitronium' has been used for the NO$_2^+$ cation; even though this cation can be named 'nitrylium' by applying Rule C-83.1 of the IUPAC *Nomenclature of Organic Chemistry*[1] to the radicals given by Recommendation I-8.4.2.2. of the IUPAC *Nomenclature of Inorganic Chemistry*[3], the name nitronium cannot be used for NH$_4^+$ without ambiguity because of this previous usage.

Examples:

CH$_3$–NH$_3^+$ (CH$_3$)$_2$OH$^+$
Methylammonium   Dimethyloxonium

(CH$_3$)$_4$N$^+$ (C$_6$H$_5$)$_2$I$^+$
Tetramethylammonium   Diphenyliodonium

A cation derived formally by the addition of one or more hydrons to any position of a neutral parent hydride is described by replacing the final 'e' of the parent hydride name, if any, by the suffix '-ium', or by adding the suffix '-ium', '-diium', etc., to the name of the parent hydride.

Examples:

CH$_5^+$
Methanium   or [C$_6$H$_7$]$^+$

[C$_2$H$_7$]$^+$ Benzenium
Ethanium

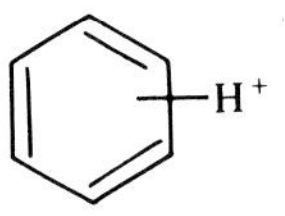

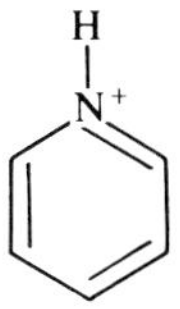

Pyridin-1-ium

$$(CH_3)_2 \overset{+}{\underset{2}{N}} = \overset{+}{\underset{1}{N}} (CH_3)_2$$

Tetramethyldiazene-1,2-diium

Cations formally derived by the removal of one hydride ion, H⁻, from a parent hydride are named using the suffix '-ylium', in the same way as the suffix '-yl' (see R-5.8.1.1). Di- and polycations formally derived by the removal of two or more hydride ions from a parent hydride are named by adding the suffixes '-bis(ylium)', '-tris(ylium)', etc., to the name of the parent hydride. A cation that can be considered to be formally derived by the loss of an unpaired electron from the corresponding radical may also be named by adding the class name 'cation' as a separate word following the name of the radical.

Examples:

CH_3^+
Methylium
Methyl cation

GeH_3^+
Germylium
Germyl cation

$CH_3-CH_2-CH_2^+$
Propylium
Propyl cation

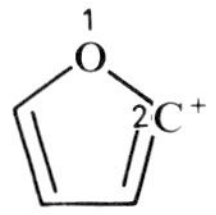

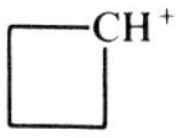

Cyclobutylium
Cyclobutyl cation

$\overset{+}{\underset{2}{C}}H_2-\overset{+}{\underset{1}{C}}H_2$
Ethane-1,2-bis(ylium)
Ethylene dication

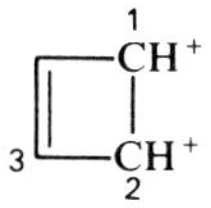

Cyclohexane with C²⁺

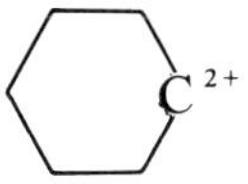

Cyclohexane-1,1-bis(ylium)
Cyclohexylidene dication

HS^+
Sulfanylium
Sulfanyl cation

$SiH_3-\overset{+}{\underset{2}{Si}}H-\underset{1}{SiH_3}$
_3
Trisilan-2-ylium
Trisilan-2-yl cation

Furan ring with C⁺ at 2-position

Furan-2-ylium
Furan-2-yl cation

$^+\underset{3}{C}H_2-\underset{2}{C}H_2-\underset{1}{C}H_2^+$
Propane-1,3-bis(ylium)
Propane-1,3-diyl dication

Cyclobutene ring with CH⁺ groups

Cyclobut-3-ene-1,2-bis(ylium)
Cyclobut-3-ene-1,2-diyl dication

$^+\underset{3}{C}H_2 - \overset{+}{\underset{2}{C}}H-\underset{1}{C}H_2^+$
Propane-1,2,3-tris(ylium)
Propane-1,2,3-triyl trication

Cations formally derived by the loss of all the hydroxy groups as hydroxide ions from acid characteristic groups expressed by the suffix are named by replacing an '-ic acid' ending or suffix by '-ylium', or a '-carboxylic acid' suffix by '-carbonylium'. Such cations may also be named by adding the class name 'cation' as a separate word after the name of the acyl group.

Examples:

$$CH_3-\overset{\overset{O}{\|}}{C}{}^+$$
Acetylium
Acetyl cation

$$C_6H_5-\overset{\overset{O}{\|}}{\underset{\|}{S}}{}^+ \\ O$$
Benzenesulfonylium
Benzenesulfonyl cation

$$C_6H_{11}\overset{+}{C}=O$$
Cyclohexanecarbonylium
Cyclohexanecarbonyl cation

Prefix names for expressing monovalent monocationic parent cations as substituents with the free valence at the cationic site are formed by changing the '-ium' or '-onium' ending of the parent cation name to '-io' or '-onio', respectively.

Examples:

H_3N^+-
Ammonio

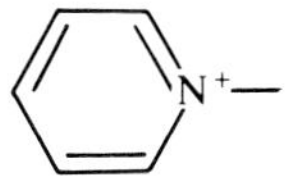

Pyridinio

Prefixes for expressing substituent structural units containing cationic centres are derived systematically by adding the operational suffixes '-yl', '-ylidene', '-diyl', etc., together with appropriate locants, to the name of the parent cation.

Examples:

$CH_3-\overset{+}{C}H_3-$
Ethan-1-ium-1-yl

$^+CH_2-CH= \\ {}_{2}{}_{1}$
Ethan-2-ylium-1-ylidene

$-\overset{+}{S}H-$
λ^4-Sulfanyliumdiyl

For replacement nomenclature, cationic replacement prefixes corresponding to '-onium' cations are formed, except for bismuth, by replacing the final 'a' of the replacement prefix for the corresponding neutral heteroatom by '-onia', for example, 'azonia-', 'oxonia-', and 'thionia-'; the cationic replacement prefix corresponding to bismuthonium is 'bismuthonia-'. Cationic replacement prefixes corresponding to '-ylium' cations are formed by replacing the final 'e' of the name of the corresponding mononuclear parent hydride by '-ylia', for example, 'azanylia' and 'boranylia'.

Example:

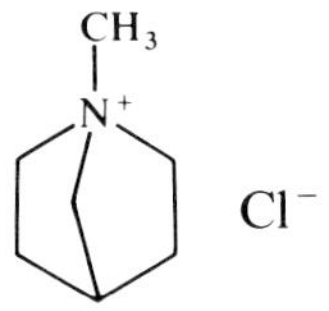

1-Methyl-1-azoniabicyclo[2.2.1]heptane chloride

R-5.8.3   **Anions**

An anion derived formally by the removal of one or more hydrons from any position of a
neutral parent hydride is named by replacing the final 'e' of the parent compound name,
if present, by the suffix '-ide', or by adding the suffixes '-diide', '-triide', etc., to the name of
the parent hydride. An anion that may be considered to be formally derived by the
addition of an electron to a radical may also be named by adding the class name 'anion'
as a separate word to the name of the radical.

Examples:

CH_3^- $CH_3-CH_2-C{\equiv}C^-$
                                                  4321
Methanide
Methyl anion   But-1-yn-1-ide
                                                  But-1-yn-1-yl anion

$(CH_3)_2CH^-$
Propan-2-ide $(C_6H_5)_2C^{2-}$
Propan-2-yl anion   Diphenylmethanediide
1-Methylethyl anion   Diphenylmethylene dianion

Note: As exceptions, the names 'amide' and 'imide' for the anions H_2N^- and HN^{2-},
respectively, are retained.

An anion formally derived by the addition of a hydride ion, H^- to a parent hydride may
be named by replacing the final 'e' of the name of the parent hydride, if present, by the
suffix '-uide'[97].

Example:
$(CH_3)_4B^-$
Tetramethylboranuide

An anion formed by the loss of the hydrogen atom as a hydron from the chalcogen atom
of an acid characteristic group or functional parent compound is named by replacing the
'-ic acid' or '-ous acid' ending of the acid name by '-ate', or '-ite', respectively.

[97] In previous recommendations[3], such anions could only be named by the principles of coordination
nomenclature, for example, tetrahydroborate(1-) for BH_4^-, or by replacement nomenclature using prefixes
ending in '-ata', for example, 'borata-' for $-\bar{B}H_2-$.

Examples:

CH₃–COO⁻
Acetate

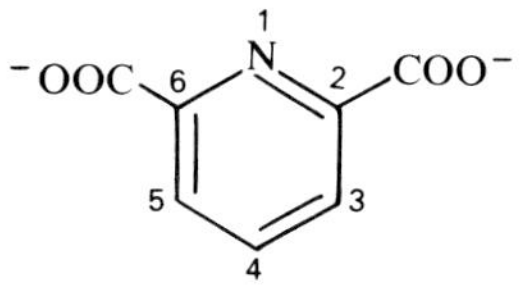

Pyridine-2,6-dicarboxylate

C₆H₅–SO₃⁻
Benzenesulfonate

(C₆H₅–CH₂)₂P–O⁻
Dibenzylphosphinite

An anion formed by the loss of the hydrogen atom as a hydron from the chalcogen atom of a hydroxy characteristic group, or a chalcogen analogue, that can be expressed as a suffix such as '-ol', '-thiol', etc., is named by adding the suffix '-ate' to the appropriate suffix. When the corresponding RO– group has a contracted name, for example, methoxy (see R-9.1, Table 26(b), p. 173), the anion name may be formed by changing the ending 'methoxy' to 'methoxide'.

Examples:

CH₃–O⁻
Methanolate
Methoxide

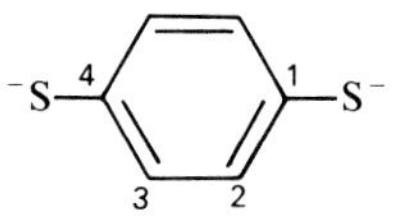

Benzene-1,4-bis(thiolate)[98]

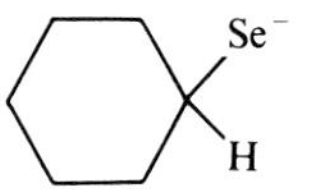

Cyclohexaneselenolate

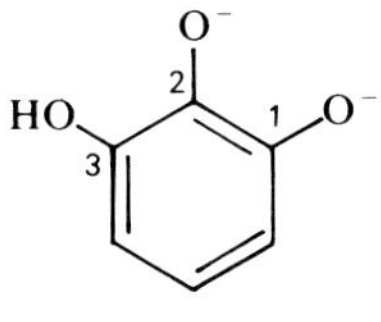

3-Hydroxybenzene-1,2-bis(olate)

Prefixes for expressing anionic centres derived from acid characteristic groups are named by changing the '-ic acid' ending of the acid suffix or of the parent mononuclear acid name to '-ato'.

Examples:

–COO⁻
Carboxylato

–PO(O⁻)₂
Phosphonato

–SO₃⁻
Sulfonato

Prefixes for expressing anionic chalcogen atoms or chains may be derived by replacing the final 'e' of the name of the anion by 'o'.

---

[98] The multiplicative prefix 'bis-' is used here because the suffix '-diolate' or '-dithiolate' could be interpreted as describing the monoanions HO–C₆H₄–O⁻ or HS–C₆H₄–S⁻.

Example:
–O⁻

Oxido

Prefixes for expressing substituent structural units containing anionic centres are derived systematically by replacing the final 'e' of the name of the parent anion with an operational suffix '-yl' or '-ylidene', or by adding an operational suffix such as '-diyl' to the name of the parent anion.

Examples:

–CH$_2$⁻   =N⁻

Methanidyl   Amidylidene

–S–S⁻

Disulfanidyl

(previously Disulfido)

For replacement nomenclature, anionic replacement prefixes are derived by replacing the final 'e' of the name of the mononuclear parent hydride (see R-2.1) by the operational suffix '-ida' for anions corresponding to '-ide' and '-uida'[99] for anions corresponding to '-uide'.

Examples:

–P⁻–   –B̄H$_2$–

Phosphanida   Boranuida

R-5-8.4   **Cationic and anionic centres in a single structure**

Zwitterionic compounds are named by combining appropriate suffixes from Table 6, p. 64, at the end of the name of a parent hydride (neutral or ionic) in the order '-ium', '-ylium', '-ide', and '-uide'.

Examples:

(CH$_3$)$_3$–N⁺–N⁻–CH$_3$ (CH$_3$)$_3$N⁺–NH–SO$_2$–O⁻
        2   1   1   2

1,2,2,2-Tetramethyldiazan-2-ium-1-ide   1,1,1-Trimethyldiazan-1-ium-2-sulfonate

1,2,2,2-Tetramethylhydrazin-2-ium-1-ide   1,1,1-Trimethylhydrazin-1-ium-2-sulfonate

R-5.8.5   **Radical ions**

A radical ion formally derived by the removal of one or more hydrogen atom(s) from a single skeletal atom or from different skeletal atoms of an ionic or zwitterionic parent hydride is named by replacing the final 'e' of its name, if present, or by adding to its name, operational suffixes, such as '-yl', '-ylium', '-diyl', etc. The λ-convention may be used when necessary.

---

[99] In previous recommendations[1], anionic replacement prefixes ending in '-ata', for example, 'borata-' for –B̄H$_2$– were used for 'uide' anions; there was no replacement prefix for 'ide' anions.

Examples:

$H_2C^{\bullet\,+}$

Methyliumyl

λ^2-Methaniumyl

$^+CH_4\text{–}CH_2^{\bullet}$
${}_2\phantom{\text{–}CH}{}_1$

Ethan-2-ium-1-yl

$^-CH_2\text{–}CH_2^{\bullet}$
${}_2\phantom{\text{–}CH}{}_1$

Ethan-2-id-1-yl

$CH_3\text{–}N{=}\overset{\bullet\,+}{N}\text{–}N^-\text{–}Si(CH_3)_3$
$\phantom{CH_3\text{–}}{}_3\phantom{N{=}}{}_2\phantom{\overset{\bullet}{N}\text{–}}{}_1$

3-Methyl-1-(trimethylsilyl)triaz-2-en-2-ium-1-id-2-yl

R-6 Name Interpretation

R-6.0 INTRODUCTION

This section provides guidelines for deducing the structure of a compound from a systematic name. For this, it is important to keep in mind that a systematic name generally consists of three main parts: (a) a prefix, or prefixes, corresponding to substituents; (b) the name of a parent hydride which includes nondetachable prefixes for heteroatoms, hydrogenation, and skeletal modification, or the name of a functional parent compound; and (c) in the case of a name based on a parent hydride, an infix, or infixes, indicating unsaturation, followed by a suffix indicating the nature of the principal characteristic group.

In the following examples, infixes and suffixes are doubly underlined and the names of parent hydrides are singly underlined; prefixes are not underlined.

R-6.1 6-(4-HYDROXYHEX-1-EN-1-YL)UNDECA-2,4-DIENE-7,9-DIYNE-1,11-DIOL

(a) The term 'undeca' in this name indicates the presence of the 11-carbon saturated parent hydride undecane.

$$CH_3–CH_2–CH_2–CH_2–CH_2–CH_2–CH_2–CH_2–CH_2–CH_2–CH_3$$
$$\quad 11 \quad\; 10 \quad\;\; 9 \quad\;\; 8 \quad\;\; 7 \quad\;\; 6 \quad\;\; 5 \quad\;\; 4 \quad\;\; 3 \quad\;\; 2 \quad\;\; 1$$

The suffix '-ol' with its multiplicative prefix 'di-' and locants '1' and '11' indicates the presence of two hydroxy (–OH) groups at positions 1 and 11 of undecane. The infixes '-ene-' and '-yne-', with their multiplicative prefix 'di-' and locants '2' and '4', and '7' and '9', describe double bonds at positions 2 and 4 and triple bonds at positions 7 and 9, respectively.

The resulting parent structure with its numbering is:

$$HO–CH_2–C≡C–C≡C–CH_2–CH=CH–CH=CH–CH_2–OH$$
$$\quad\;\; 11 \quad\; 10 \; 9 \; 8 \; 7 \quad 6 \quad\; 5 \quad\;\; 4 \quad\; 3 \quad\; 2 \quad\;\; 1$$

(b) The complex prefix '(4-hydroxyhex-1-en-1-yl-)' with its locant '6' describes a substituent at position 6 of the parent structure. The parent substituent is a six-carbon chain corresponding to the parent hydride name 'hexane'. The prefix 'hydroxy-' with its locant '4' denotes a hydroxy (–OH) group at position 4; and the infix '-en-' with its locant '1' and the suffix '-yl-' with its locant '1' describe, respectively, a double bond at position 1 and a free valence at position 1. Hence, the structure and numbering of the complex substituent is:

$$\overset{\textstyle OH}{\underset{\textstyle |}{}}$$
$$CH_3–CH_2–CH–CH_2–CH=CH—$$
$$\;\; 6 \quad\;\; 5 \quad\; 4 \quad\;\; 3 \quad\; 2 \quad\; 1$$

143

(c) The structure of the complete compound is therefore:

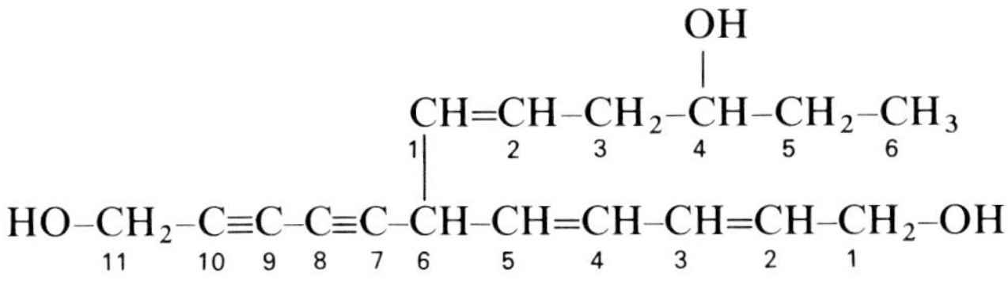

$$
\begin{array}{c}
\text{OH} \\
\text{CH=CH–CH}_2\text{–CH–CH}_2\text{–CH}_3 \\
\overset{1}{}\quad\overset{2}{}\quad\overset{3}{}\quad\overset{4}{}\quad\overset{5}{}\quad\overset{6}{}
\end{array}
$$

$$
\text{HO–CH}_2\text{–C}\equiv\text{C–C}\equiv\text{C–CH–CH=CH–CH=CH–CH}_2\text{–OH}
$$
$$
11\quad10\quad9\quad8\quad7\quad6\quad5\quad4\quad3\quad2\quad1
$$

2,3-DICHLORO-6-[4-CHLORO-2-(HYDROXYMETHYL)-5-OXOHEX-3-EN-1-YL]PYRIDINE-4-CARBOXYLIC ACID

(a) The suffix carboxylic acid together with its locant, '4', describes the presence of a carboxy group, – COOH, at position 4 of the parent hydride 'pyridine' giving the structure and numbering shown below:

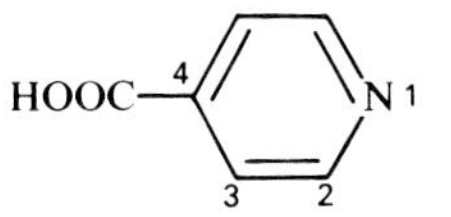

(I)

(b) The prefixes 'chloro-' and '[4-chloro-2-(hydroxymethyl)-5-oxohex-3-en-1-yl-]', together with the numerical prefix 'di-' and the locants '2, 3', and '6', respectively, indicate the presence of chlorine atoms at positions 2 and 3 and a complex substituent 'R' at position 6; the resulting partial structure is as follows:

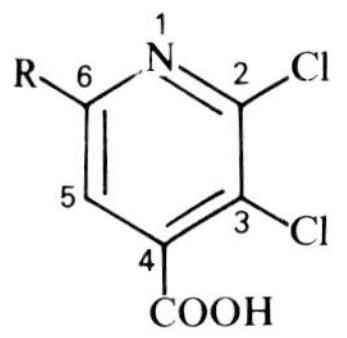

(II)

(c) The complex substituent 'R' is analysed in the same way as in (a) and (b), except that the point of attachment (free valence) takes the place of the principal characteristic group. Hence, in the substituent name 4-chloro-2-(hydroxymethyl)-5-oxohex-3-en-1-yl:

    (1) the suffix '-yl', together with its locant '1', specifies the attachment of a six-carbon chain, hex, to the parent structure (II); and the subtractive infix 'en', together with its locant, '3', indicates the presence of a double bond between chain atoms 3 and 4, giving the following structure:

$$
\underset{6}{\text{CH}_3}\text{–}\underset{5}{\text{CH}_2}\text{–}\underset{4}{\text{CH}}\text{=}\underset{3}{\text{CH}}\text{–}\underset{2}{\text{CH}_2}\text{–}\underset{1}{\text{CH}_2}\text{–}
$$
                                     (III)

    (2) the prefixes 'chloro-', 'hydroxymethyl-', and 'oxo-' together with their respective locants '4', '2', and '5', indicate the presence of a chlorine atom, a hydroxymethyl group, and a doubly-bonded oxygen atom substituting hydrogen atoms at these positions of the parent substituent, giving the structure shown as (IV).

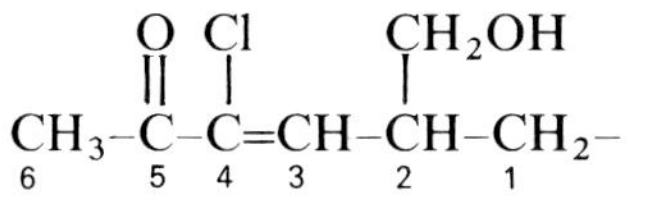

$$
\begin{array}{c}
\quad\quad\text{O} \;\; \text{Cl} \quad\quad \text{CH}_2\text{OH} \\
\quad\quad \| \;\; | \quad\quad\quad\; | \\
\underset{6}{\text{CH}_3}\text{–}\underset{5}{\text{C}}\text{–}\underset{4}{\text{C}}\text{=}\underset{3}{\text{CH}}\text{–}\underset{2}{\text{CH}}\text{–}\underset{1}{\text{CH}_2}\text{–}
\end{array}
\quad\quad\text{(IV)}
$$

144

(d) Attaching 'R', shown in structure (IV), to structure (II) gives the complete structure of the compound, as follows:

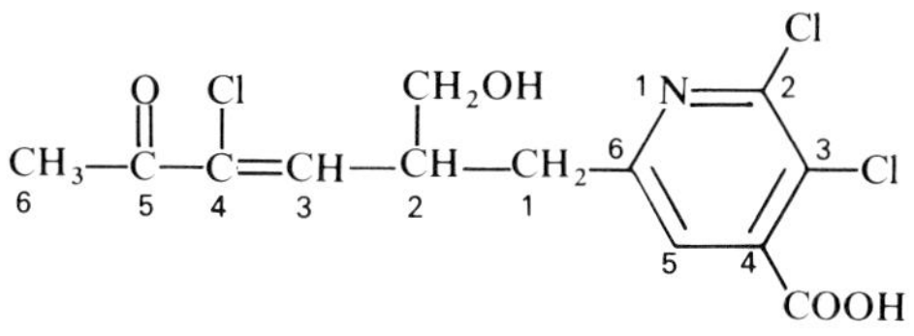

3-(2,3-DIHYDROXYPROPYL)-α-METHYLQUINOLINE-2-PENTANOIC ACID

The name of the parent structure in this example is a conjunctive name formed by the juxtaposition of the names of the two components: (1) an acyclic chain containing the principal group, pentanoic acid; and (2) a heterocycle, quinoline. It is implied that each component has lost one hydrogen atom to form the bond (see R-1.2.4).

(a) The suffix '-oic acid' denotes the presence of a carboxy [–(C)OOH] group attached to the parent hydride 'pentane' (see R-5.7.1), giving the structure and numbering shown as (I), below. Note that the carbon atom of the carboxy group is included in the parent chain and that the numbering of the chain, although shown, is not that used to indicate substituents of the chain in the name of the complete compound. The structure and numbering of the heterocyclic component, quinoline, is given as (II).

$$CH_3-CH_2-CH_2-CH_2-COOH \qquad \text{(I)}$$
$$54321$$

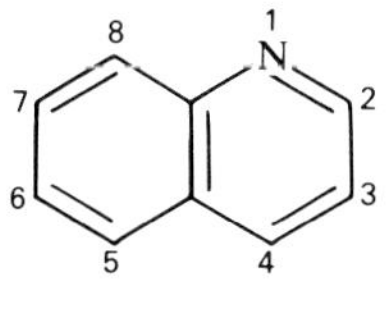

(II)

The locant '2' is the position on the heterocyclic component, quinoline, at which the acyclic component, pentanoic acid, is attached. No locant is given to denote the position of the acyclic component attached to the heterocycle, because, by definition, the acyclic component must be attached at the atom furthest from the principal group (see R-1.2.4.1). The structure and numbering of the conjunctive parent structure, quinoline-2-pentanoic acid is shown as (III).

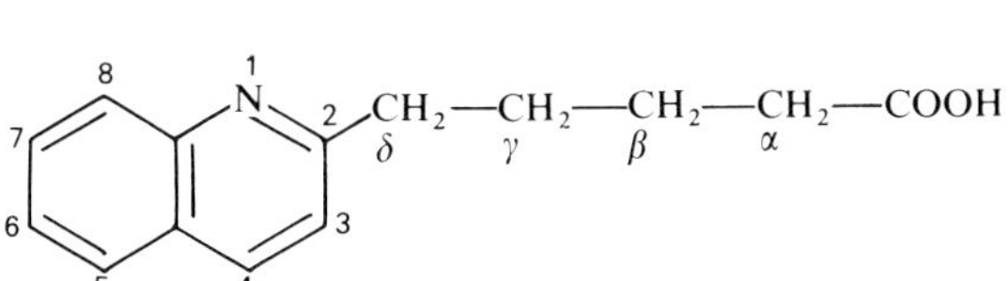

(III)

(b) The prefixes '(2,3-dihydroxypropyl)-' and 'methyl-', together with the locants '3' and 'α', respectively, indicate the presence of these groups as substituents at positions '3' and 'α' of the conjunctive parent structure.

(1) The prefix 'hydroxy-', the multiplicative prefix 'di-' and the locants '2' and '3', denote the presence of two hydroxy groups at positions 2 and 3 of a three-carbon chain (prop), derived from the parent hydride propane, with its free valence at position 1, as shown in (IV).

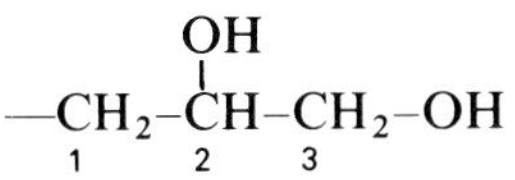

$$-\underset{1}{CH_2}-\underset{2}{\overset{OH}{\underset{|}{CH}}}-\underset{3}{CH_2}-OH \qquad (IV)$$

(c) Attaching structure (IV) and the methyl group to the conjunctive parent structure (III) completes the structure of the compound shown below.

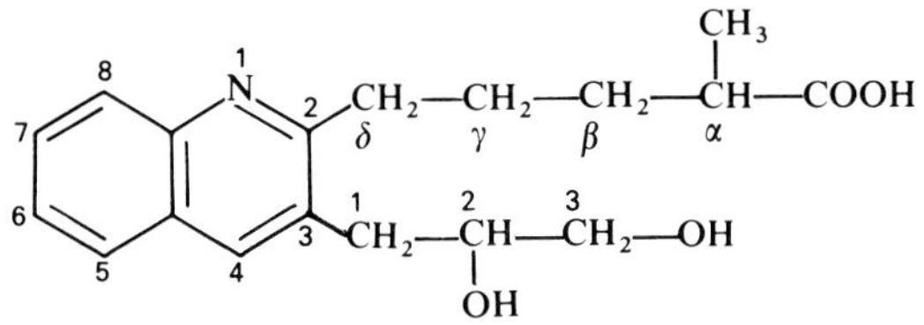

R-6.4   4,4'-DINITRO-2,3'-[ETHYLENEBIS(SULFANEDIYL)]DICYCLOHEXANE-1-CARBALDEHYDE

This example illustrates a name involving an assembly of identical units.

(a) The suffix '-carbaldehyde' indicates the presence of the principal group –CHO attached to the parent hydride 'cyclohexane' giving the structure and numbering of cyclohexanecarbaldehyde shown as (I), below. The multiplicative prefix 'di-' denotes the presence of two identical units, and the locants '2,3'' indicate attachment of a bivalent linking group –Y– at position 2 of one cyclohexane ring and position 3' of a second cyclohexane ring (priming of locants differentiates two or more identical units of an assembly). The structure and numbering of the resulting partial structure is shown as (II).

(I) 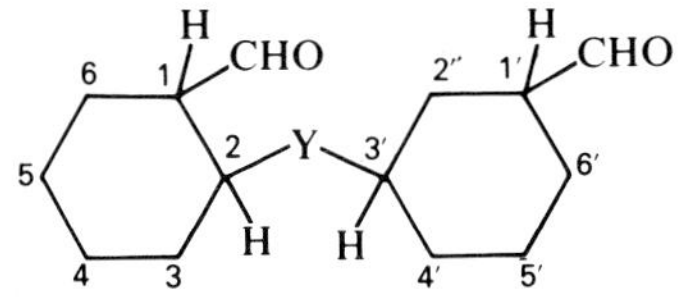   (II)

(b) The prefix 'ethylenebis(sulfanediyl)' denotes a bivalent linking group made up of a central ethylene group ($-CH_2-CH_2-$) attached to two sulfur atoms, thus, $-S-CH_2-CH_2-S-$.

(c) The prefix 'nitro-', with its multiplicative prefix 'di-' and the locants '4,4″', indicates the presence of two nitro groups, $-NO_2$, one attached to position 4 of one cyclohexane ring and one to position 4′ of the other.

(d) The structure of the complete compound is therefore:

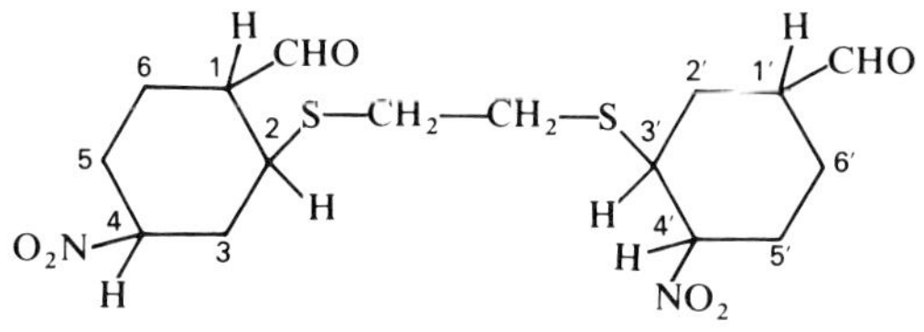

1-METHYLBUTYL 4-(2-ACETYL-2-ETHYLHYDRAZINO)BENZOATE

This example illustrates a functional class name, and is composed of two separate words describing an ester Ar–CO–OR derived from an acyclic alcohol, R-OH, and an arenecarboxylic acid, Ar–COOH.

(a) The acyclic alcohol from which this ester is derived is described by the first word '1-methylbutyl'. The prefix 'butyl' indicates a four-carbon chain derived from the parent hydride 'butane'. The prefix 'methyl-' with its locant '1' indicates that a methyl group is at position '1' of the butyl group which, by definition, also bears the free valence, resulting in the following structure and numbering for the alcoholic component:

$$\underset{4}{CH_3}-\underset{3}{CH_2}-\underset{2}{CH_2}-\underset{1}{\overset{\overset{\displaystyle CH_3}{|}}{CH}}—$$

(b) The acid from which this ester is derived is described by the second word. The term 'benzoate' is derived from the functional parent compound benzoic acid (I). The suffix '-oate' can describe either an ester or an anion; here, it describes an ester.

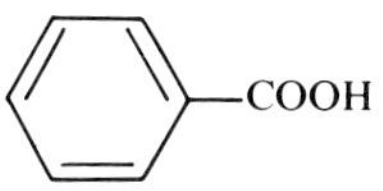

                                   (I)

(1) The prefix '(2-acetyl-2-ethylhydrazino)-' together with the locant '4' indicates the presence of a substituent at position 4 of benzoic acid. The prefixes 'acetyl-' and 'ethyl-' with the same locant '2' indicate the presence of the groups 'acetyl' (CH_3–CO–) and 'ethyl' (CH_3–CH_2–) at position 2 on the parent substituent 'hydrazino', derived from the parent hydride hydrazine. By definition, the free valence for the hydrazino

substituent must have the locant '1'; thus, the structure and numbering of the substituent of the acid component is shown as (II). The structure and numbering for the acid component of this ester are given as (III).

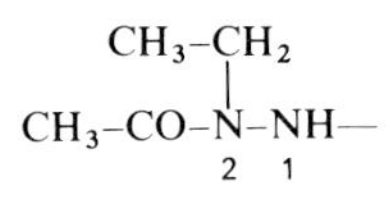

CH$_3$–CH$_2$
|
CH$_3$–CO–N–NH—
    2 1

(II)

CH$_3$—CH$_2$
|
CH$_3$—CO—N—NH-4⟨ ⟩1-COOH
      2 1   3 2

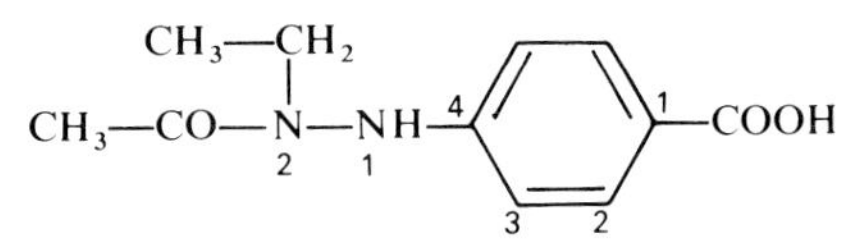

(III)

(c) The structure of the complete ester is therefore:

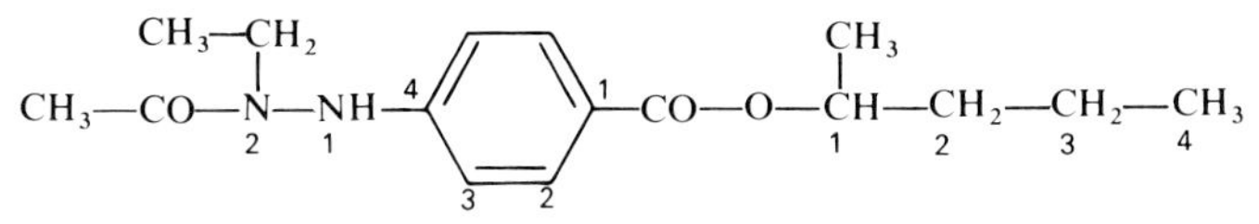

CH$_3$—CH$_2$ CH$_3$
| |
CH$_3$—CO—N—NH-4⟨ ⟩1-CO—O—CH—CH$_2$—CH$_2$—CH$_3$
        2 1   3 2   1   2   3   4

R-7   Stereochemical Specification

R-7.0   INTRODUCTION

This section is concerned only with the main principles for specification of stereochemistry in organic compounds [for a more detailed description, see Section E of the IUPAC *Nomenclature of Organic Chemistry*[1]; see also *Pure Appl. Chem.*, **45**, 11–30 (1976)]. The spatial structure of an organic compound is systematically indicated by one or more affixes added to a name that does not itself prescribe stereochemical configuration; such affixes are generally called stereodescriptors and do not change the name or the numbering of a compound established by the principles of nomenclature described in other sections of these recommendations. Thus, stereoisomers, such as enantiomers and *cis/trans* isomers, have names that differ only in the stereodescriptors used. By contrast, certain trivial names imply their own stereochemical configuration, for example, fumaric acid and cholesterol.

R-7.1   *cis-trans* ISOMERISM – THE *E/Z* CONVENTION

R-7.1.0   **Introduction**

Stereoisomers that differ only in the positions of atoms relative to a specified plane in cases where these atoms are, or are considered as if they were, parts of a rigid structure are specified by the use of the stereodescriptors *cis* and *trans*[100] or *E* and *Z*.

Note: The Greek letters α and β are also used as stereodescriptors in specific classes of compounds, such as steroids[101].

R-7.1.1   *cis* **and** *trans* **Isomers**

Atoms or groups are called *cis* or *trans* to one another when they project respectively on the same or on opposite sides of a reference plane identifiable as common among stereoisomers. The compounds in which such relations occur are termed *cis/trans*-isomers. For doubly bonded atoms, the reference plane contains these atoms and is perpendicular to the plane containing the doubly bonded atoms and the atoms directly attached to them. For cyclic compounds, the reference plane is that in which the ring skeleton lies or to which it approximates. When alternative numberings of a ring or ring system are possible, that numbering is chosen which gives a *cis* attachment at the first point of difference.

---

[100] *cis*- and *trans*- may be abbreviated to *c*- and *t*-, respectively, in names when more than one such designation is required.

[101] When the rings of a steroid are represented as projections onto a plane in the conventional manner, an atom or group is called α if it projects below the plane and β if it projects above the plane. This 'α/β-system' has been extended to other cyclic natural products and related products (see, for instance, Section E, Stereochemistry and Stereoparents, pp. 192I–212I, in 'Chemical Substance Index Names', Appendix IV, in the 1992 *Index Guide*, Chemical Abstracts Services, Columbus, Ohio 1992).

Examples:

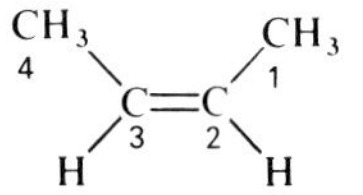

cis-But-2-ene[102]

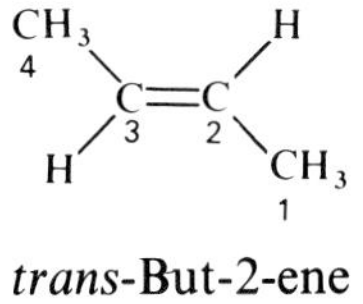

trans-But-2-ene

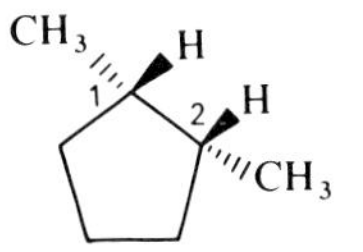

cis-1,2-Dimethylcyclopentane[103]

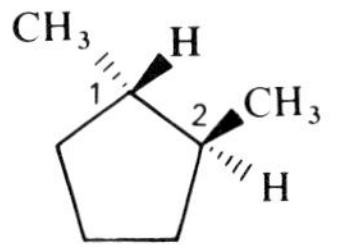

trans-1,2-Dimethylcyclopentane

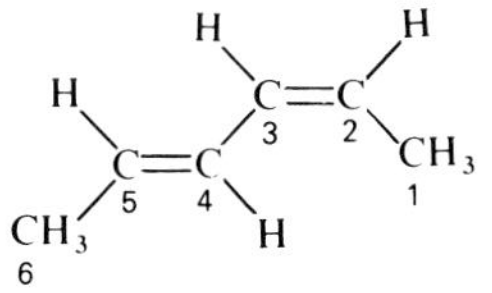

2-cis,4-trans-Hexa-2,4-diene

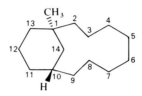

1-Methyl-trans-bicyclo[8.3.1]tetradecane

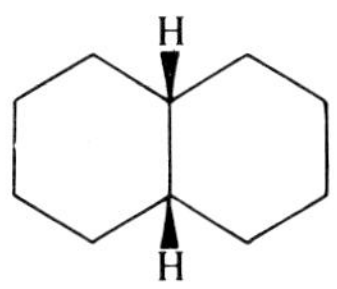

cis-Decahydronaphthalene

When one substitutent and one hydrogen atom are attached at each of more than two positions of a monocycle, the steric relations of the substituents are expressed by adding *r* (for reference substituent) followed by a hyphen and the locant of the lowest numbered of these substituents and *c* and *t* (as appropriate) followed by a hyphen and the locant of another substituent, thus expressing its relationship to the reference substituent.

Examples:

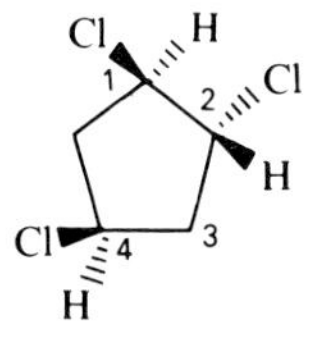

r-1,t-2,c-4-Trichlorocyclopentane

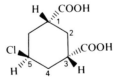

t-5-Chlorocyclohexane-r-1,c-3-dicarboxylic acid

When two different substituents are attached at the same position of a monocycle, then the lowest-numbered substituent named as a suffix is selected as reference group. If none of the substituents is named as a suffix, then that substituent of the pair of substituents

---

[102] The *E/Z*-convention is preferred for describing geometrical isomerism about double bonds (see R-7.1.2).
[103] By convention, a broken wedged line denotes a bond projecting behind the plane of the paper and a solid wedged line denotes a bond projecting in front of the plane; a normal line denotes a bond lying in the plane.

having the lowest number, and which is preferred by the sequence rule (see R-7.2.1), is chosen as the reference group. The relationship of the sequence-rule-preferred substituent at geminally substituted positions, relative to the reference group, is cited as c- or t-, as appropriate.

Examples:

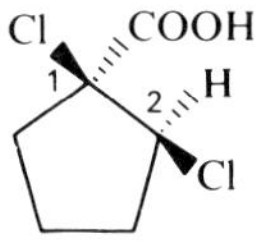

1,t-2-Dichlorocyclopentane-r-1-
  carboxylic acid

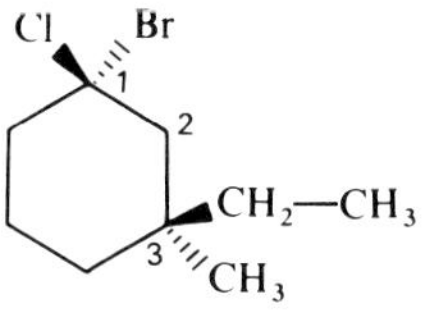

r-1-Bromo-1-chloro-t-3-ethyl-3-
  methylcyclohexane

R-7.1.2   **The *E/Z* convention**

In names of compounds, steric relationships around one or more double bonds can be designated by the stereodescriptors *Z* and/or *E*[104], assigned as follows. The sequence-rule-preferred atom or group (see R-7.2.1) attached to one of a doubly bonded pair of atoms is compared with the sequence-rule-preferred atom or group attached to the other of that doubly bonded pair of atoms; if the selected atoms are on the same side of the reference plane (see Rule R-7.1.1) the italic capital letter *Z* is used as the stereodescriptor; if the selected atoms are on opposite sides, the italic capital letter *E* is used as the stereodescriptor. These stereodescriptors, placed in parentheses followed by a hyphen, normally precede the whole name; if the molecule contains several double bonds, then each stereodescriptor is immediately preceded by the lower or less primed locant of the relevant double bond.

Examples:

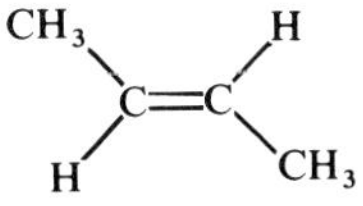

(*E*)-But-2-ene

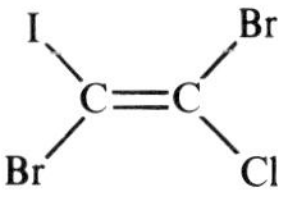

(*Z*)-1,2-Dibromo-1-chloro-2-iodoethene
  (By the sequence rule, Br is preferred
    to Cl, but I to Br).

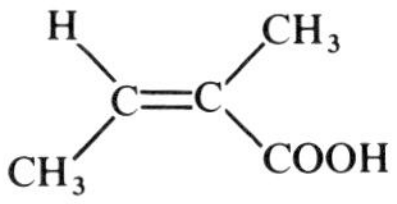

(*Z*)-2-Methylbut-2-enoic acid

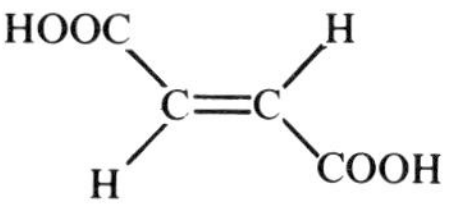

(*E*)-But-2-enedioic acid
  Fumaric acid (the stereochemistry is
    implied in this trivial name)

---

[104] These affixes have been rationalized as derived from the German words 'Zusammen' (together) and 'Entgegen' (opposite).

(E)-2,3-Dichloroacrylonitrile   (2E,4Z)-Hexa-2,4-dienoic acid

R-7.2   CHIRAL COMPOUNDS – SPECIFICATION OF
               ABSOLUTE CONFIGURATION

R-7.2.1   **The R/S convention**
               Chiral compounds in which the absolute configuration is known are differentiated by the
               stereodescriptors R and S assigned by the sequence-rule procedure (see also Section E in
               the 1979 edition of the IUPAC *Nomenclature of Organic Chemistry*[1], p. 486 and R. S.
               Cahn, C. K. Ingold, and V. Prelog, *Angew. Chem.*, **78**, 413–447 (1966); and V. Prelog and
               G. Helmchen, *Angew. Chem., Int. Ed.*, **21**, 567–583 (1982)) and are preceded when
               necessary by appropriate locants. For carbon (or other atoms) to which four different
               ligands are attached in the system Cabcd may be depicted as in **1**, where a > b > c
               > d[105].

               1 (R)   **2** (S)

               If the model **1** is viewed from the side remote from the fourth member d, the path from a
               to b to c traces a clockwise direction, and the system has the 'R' configuration; if passing
               from a to b to c traces an anticlockwise direction, as in **2**, the configuration is 'S'.
               The sequence rule itself is the method by which the groups a, b, c, d are assigned
               priorities. It contains five sub-rules, but only the first two are mentioned here, the other
               sub-rules being needed in relatively few cases.

               Sub-rule 1: Higher atomic number precedes lower.

               Here, the atomic numbers of the atoms Br, Cl, C, H attached to the central carbon atom
               are arranged in that order by sub-rule 1, giving the model shown, which is in the R
               configuration.

---

[105] The symbol > here means 'precedes' in terms of sequence-rule priority.

When two or more atoms attached to the chiral centre are alike, the comparison is extended through each of these atoms to succeeding atoms until a sequence priority can be established on the same basis. Multiple bonds are considered as two or three single bonds to the same atom. Thus, a $-CH=O$ group is treated as

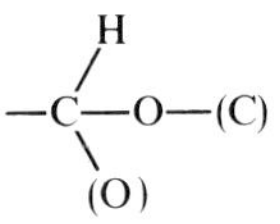

$$-\overset{\displaystyle H}{\underset{\displaystyle (O)}{C}}-O-(C) \qquad\text{and } -CH=CH_2 \text{ as} \qquad -\overset{\displaystyle H}{\underset{\displaystyle (C)}{C}}-\overset{\displaystyle H}{\underset{\displaystyle H}{C}}-(C)$$

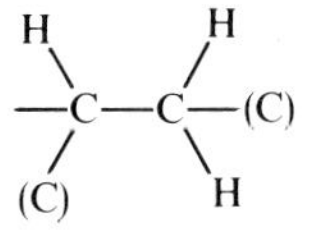

where (O) and (C) are 'duplicate representations' of the respective atoms which, if needed for further comparison, are attached only to 'phantom atoms' having neither atomic number nor mass. Since O is preferred to C when the sequences O, (O), H and C, (C), H are compared, the $-CHO$ group has sequence rule priority over the $-CH=CH_2$.

Example:

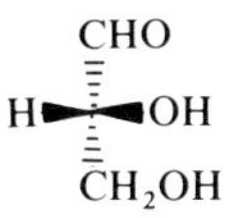

          CHO   CHO
H◀―――OH   HO◀―――H
      CH$_2$OH   CH$_2$OH

(*R*)-Glyceraldehyde[106] 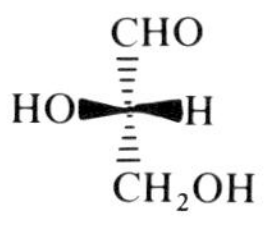   (*S*)-Glyceraldehyde

Similarly, a $-C\equiv N$ group is treated as

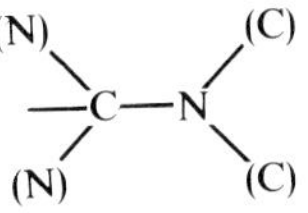

$$-\overset{\displaystyle (N)}{\underset{\displaystyle (N)}{C}}-\overset{\displaystyle (C)}{\underset{\displaystyle (C)}{N}}$$

Sub-rule 2: Higher mass number precedes lower (applies only when sub-rule 1 does not decide priority).

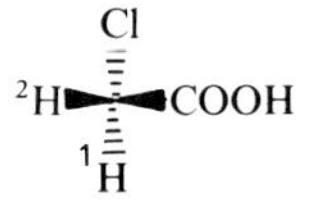

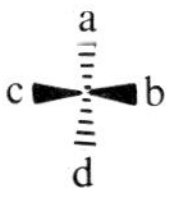

          Cl   a
^{2}H▶―――COOH   c▶―――b
      ^{1}H   d

                    R

In this case, the mass numbers allow the atoms to be arranged in the order Cl, C, ^{2}H, ^{1}H, resulting in an *R* configuration.

If a molecule contains more than one chiral centre, this procedure is applied to each, and the configuration is expressed as a set of *R,S* symbols. In names of compounds, the symbols *R* and *S*, with locants if needed, are written in parentheses, followed by a hyphen, in front of the complete name or appropriate substituent.

[106] D and L have been used to characterize (*R*)- and (*S*)-glyceraldehyde, respectively.

Examples:

(5′R,6aS,12aS)-Rotenone

(S)-Methyl phenyl sulfoxide
(Here the lone pair of electrons is
    considered as the 'substituent' of
    lowest priority).

R-7.2.2   **Relative configuration**

Chiral centres, of which the relative but not the absolute configuration is known, may be labelled arbitrarily as *R** or *S** (spoken R star, S star), preceded when necessary by appropriate locants. These prefixes are assigned by the sequence rule procedure, arbitrarily assuming that the centre of chirality first cited (which usually has the lowest locant) has the chirality descriptor *R**. In complex cases, the stars may be omitted, and, instead, the whole name prefixed by *rel-* (for *relative*).

Examples:

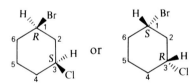

(1*R**,3*S**)-1-Bromo-3-chlorocyclohexane

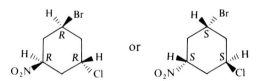

rel-(1*R*,3*R*,5*R*)-1-Bromo-3-chloro-5-nitrocyclohexane

R-8   Isotopically Modified Compounds

R-8.0   INTRODUCTION

This section describes a general system of nomenclature for organic compounds whose isotopic nuclide[107] composition deviates from that occurring in nature[108]. Comparative examples of the application of these rules are given in Table 18, p. 161. Additional details may be found in Section H in the 1979 edition of the IUPAC *Nomenclature of Organic Chemistry*[1].

There is one other general system in use for describing isotopically modified compounds. It is based on an extension of the principles proposed by Boughton[109] for designating compounds containing hydrogen isotopes and is currently in use mainly in the Chemical Abstracts Service index nomenclature system[110].

The system codified in these recommendations provides for recognition of various types of isotopic modification and thus was chosen in preference to the system based on the Boughton principles.

R-8.1   SYMBOLS AND DEFINITIONS

R-8.1.1   **Nuclide symbols**

The symbol for denoting a nuclide in the formula or name of an isotopically modified compound consists of the atomic symbol for the element and an arabic numeral in the left superscript position which indicates the mass number of the nuclide[111].

R-8.1.2   **Atomic symbols**

The atomic symbols used in the nuclide symbol are those given in IUPAC *Nomenclature of Inorganic Chemistry*[3]. In the nuclide symbol, the atomic symbol is printed in roman type, italicized atomic symbols being reserved for letter locants, as is customary in organic chemical nomenclature.

---

[107] International Union of Pure and Applied Chemistry, Physical Chemistry Division. Commission on Symbols, Quantities, and Units, '*Quantities, Units, and Symbols in Physical Chemistry*', Blackwell Scientific Publications, Oxford, 1988, p. 39 (commonly known as the IUPAC Green Book).

[108] For a discussion of the meaning of 'natural composition', see *Pure Appl. Chem.*, **37**, 591–603(1974). In any context where the accuracy requires it, the natural nuclide composition used should be stated.

[109] W.A. Boughton, 'Naming Hydrogen Isotopes', *Science*, **79**, 159–160(1934); E. J. Crane, 'Nomenclature of Hydrogen Isotopes and Their Compounds', *Science*, **80**, 86 89(1934); American Chemical Society, 'Report of Committee on Nomenclature, Spelling, and Pronunciation. Nomenclature of the Hydrogen Isotopes and Their Compounds', *Ind. Eng. Chem. (News Ed.)*, **13**, 200–201(1935).

[110] American Chemical Society. Chemical Abstracts Service, 'Chemical Substance Index Names', Appendix IV, *Chemical Abstracts 1992 Index Guide*, ¶220, pp. 223I–225I.

[111] See IUPAC *Nomenclature of Inorganic Chemistry*[3], I-3.2.4 and I-3.5.2, pp. 35–38.

155

Note: For the hydrogen isotopes protium, deuterium, and tritium[112], the nuclide symbols ^1H, ^2H, and ^3H, respectively, are used. The symbols D and T for ^2H and ^3H, respectively, may be used, but not when other modifying nuclides are also present, because this may cause difficulties in alphabetic ordering of the nuclide symbols in the isotopic descriptor. Although the symbols *d* and *t* have been and are still used in place of ^2H and ^3H in names formed according to the Boughton system (see R-8.0), in no other cases are lower-case letters used as atomic symbols. Therefore, the use of *d* and *t* in chemical nomenclature outside of the Boughton system is not recommended.

R-8.1.3   **Isotopically unmodified compounds**

An isotopically *unmodified* compound has a macroscopic composition such that its constituent nuclides are present in the proportions occurring in nature. Its formula and name are written in the customary manner.

Examples:

CH_4 CH_3-CH_2-OH

Methane   Ethanol

R-8.1.4   **Isotopically modified compounds**

An isotopically *modified* compound has a macroscopic composition such that the isotopic ratio of nuclides for at least one element deviates measurably from that occurring in nature. Isotopically modified compounds may be classified as isotopically *substituted* (R-8.2) or as isotopically *labelled* (R-8.3). Isotopically labelled compounds may be further classified as specifically labelled (R-8.3.1), selectively labelled (R-8.3.2), nonselectively labelled (R-8.3.3), or isotopically deficient (R-8.3.4).

R-8.2   ISOTOPICALLY SUBSTITUTED COMPOUNDS

An isotopically substituted compound has a composition such that essentially all the molecules of the compound have *only* the indicated nuclide at each designated position. For all other positions, the absence of nuclide indication means that the nuclide composition is the natural one.

R-8.2.1   **Formulae**

The *formula* of an isotopically *substituted* compound is written as usual except that appropriate nuclide symbols are used. When different isotopes of the same element are present at the same position, common usage is to write their symbols in order of increasing mass number, i.e. CH_3-CH^2H-OH, not CH_3-C^2HH-OH.

---

[112] According to the recommendations of IUPAC Commission on Physical Organic Chemistry [*Pure Appl. Chem.*, **60**, 1115–1116 (1988)], the names for hydrogen atoms and ions are the following:

		^1H	^2H	^3H	H^a
atom	H	protium	deuterium	tritium	hydrogen
cation	H^+	proton	deuteron	triton	hydron
anion	H^-	protide	deuteride	tritide	hydride

[a] in natural or unspecified isotopic abundance.

Names

The *name* of an isotopically *substituted* compound is formed by inserting in parentheses the nuclide symbol(s) preceded by any necessary locants before the name or preferably before the name of that part of the compound that is isotopically modified.

Examples:

$^{13}CH_4$ $CH_3–CH^2H–OH$
(^{13}C)Methane   (1-$^2H_1$)Ethanol

$^{12}CHCl_3$
(^{12}C)Chloroform

R-8.3   ISOTOPICALLY LABELLED COMPOUNDS

An isotopically *labelled* compound is a mixture of an isotopically unmodified compound with one or more analogous isotopically substituted compound(s).

Note: Although an isotopically labelled compound is really a mixture as far as chemical identity is concerned (in the same way as is an unmodified compound), for nomenclature purposes, such mixtures are called 'isotopically labelled' compounds.

R-8.3.1   **Specifically labelled compounds**

An isotopically labelled compound is designated as *specifically labelled* when a *unique* isotopically substituted compound is formally added to the analogous isotopically unmodified compound. In such a case, both position(s) and number of each labelling nuclide are defined.

R-8.3.1.1   The ***structural formula*** of a specifically labelled compound is written in the usual way, but with the appropriate nuclide symbol(s) and multiplying subscript, if any, enclosed in *square brackets*. Other principles used in writing the formula are described in Section H of the 1979 edition of the IUPAC *Nomenclature of Organic Chemistry*[1], page 514.

Examples:

Isotopically substituted compound	when added to	Isotopically unmodified compound	gives rise to	Specifically labelled compound
$^{13}CH_4$		CH_4		$[^{13}C]H_4$
$CH_2{}^2H_2$		CH_4		$CH_2[^2H_2]$

Note: Although the formula for a specifically labelled compound does not represent the composition of the bulk material, which usually consists overwhelmingly of the isotopically unmodified compound, it does indicate the presence of the compound of chief interest, the isotopically substituted compound.

A specifically labelled compound is (a) *singly labelled* when the isotopically substituted compound has only one isotopically modified atom, for example, $CH_3–CH[^2H]–OH$; (b) *multiply labelled* when the isotopically substituted compound has more than one modified atom of the same element at the same position or at different positions, for example, $CH_3–C[^3H_2]–OH$ and $CH_2[^2H]–CH[^2H]–OH$; or (c) *mixed labelled* when the isotopically substituted compound has more than one kind of modified atom, for example $CH_3-CH_2-[^{18}O][^2H]$.

R-8.3.1.2   The **name** of a specifically labelled compound is formed by inserting in *square brackets* the nuclide symbol(s) preceded by any necessary locants before the name or preferably before the denomination for that part of the compound that is isotopically modified. When polylabelling is possible, the number of atoms that have been labelled is always specified as a subscript to the atomic symbol(s) even in case of monolabelling. This is necessary in order to distinguish between a specifically and a selectively or nonselectively labelled compound.

Examples:

$[^{13}C]H_4$
$[^{13}C]$Methane

$CH_3[^2H]$
$[^2H_1]$Methane

$C[^2H_2]Cl_2$
Dichloro$[^2H_2]$methane

1-(Amino$[^{14}C]$methyl)cyclopentanol

R-8.3.2   **Selectively labelled compounds**

An isotopically labelled compound is designated as *selectively labelled* when a mixture of isotopically substituted compounds is formally added to the analogous isotopically unmodified compound in such a way that the position(s) but not necessarily the number of each labelling nuclide is defined. A selectively labelled compound may be considered as a mixture of specifically labelled compounds.

A selectively labelled compound may be (a) *multiply labelled* when in the unmodified compound there is more than one atom of the same element at the position where the isotopic modification occurs, for example, H in CH_4; or there are several atoms of the same element at different positions where the isotopic modification occurs, for example, C in C_4H_8O; or (b) *mixed labelled* when there is more than one labelling nuclide in the compound, for example, C and O in CH_3-CH_2-OH.

Note: When there is only one atom of an element that can be modified in a compound, only specific labelling can result.

R-8.3.2.1   A *selectively* labelled compound cannot be described by a unique structural *formula*; therefore it is represented by inserting the nuclide symbols preceded by any necessary locant(s) (letters and/or numbers) but without multiplying subscripts, enclosed in square brackets directly before the usual formula or, if necessary, before parts of the formula that have an independent numbering. Identical locants are not repeated. When different nuclides are present, the nuclide symbols are written in alphabetical order according to their symbols, or when the atomic symbols are identical, in order of increasing mass number.

Example:

Mixture of isotopically substituted compounds	when added to	Isotopically unmodified compound	gives rise to	Selectively labelled compound
$CH_3{}^2H$, $CH_2{}^2H_2$ CH^2H_3, C^2H_4 or any two or more of the above		CH_4		$[^2H]CH_4$

Note: The method of writing formulae as given by the above rule may also be of use if a compound is represented by its molecular formula rather than its structural formula, e.g., $[^2H]C_2H_6O$.

R-8.3.2.2   The **name** of a selectively labelled compound is formed in the same way as the name of a specifically labelled compound, except that the multiplying subscripts following the atomic symbols are generally omitted. Identical locants corresponding to the same element are not repeated. The name of a selectively labelled compound differs from the name of the corresponding isotopically substituted compound in the use of *square brackets* surrounding the nuclide descriptor rather than parentheses and in the omission of repeated identical locants and multiplying subscripts.

Examples:

Mixture of isotopically substituted compounds	when added to	is named
$CH_3{}^2H$, $CH_2{}^2H_2$ CH^2H_3, C^2H_4	CH_4	$[^2H]$Methane **not** $[^2H_4]$Methane
$CH_3–CH^2H–OH$ $CH_3–C^2H_2–OH$	$CH_3–CH_2–OH$	$[1-^2H]$Ethanol **not** $[1,1-^2H_2]$Ethanol

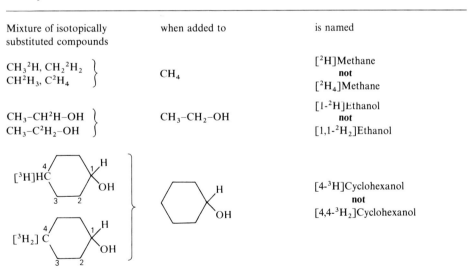

		$[4-^3H]$Cyclohexanol **not** $[4,4-^3H_2]$Cyclohexanol

In a selectively labelled compound formally arising from mixing several *known* isotopically substituted compounds with the analogous isotopically unmodified compound, the number or the possible number of labelling nuclide(s) for each position may be indicated by subscripts to the atomic symbol(s). Two or more subscripts referring to the same nuclide symbol are separated by a semicolon. For a multiply labelled or mixed labelled compound, the subscripts are written successively in the same order as the various isotopically substituted compounds are considered. The subscript zero is used to indicate that one of the isotopically substituted compounds is not modified at the indicated position.

Examples:

Mixture of isotopically substituted compounds	when added to	Isotopically unmodified compound	gives rise to	Selectively labelled compound
$CH_2{}^2H-CH_2-OH$ $CH^2H_2-CH_2-OH$		CH_3-CH_2-OH		$[2-{}^2H_{1;2}]CH_3-CH_2-OH$ $[2-{}^2H_{1;2}]$Ethanol
$CH^2H_2-CH_2-OH$ $CH^2H_2-CH_2-{}^{18}OH$		CH_3-CH_2-OH		$[2-{}^2H_{2;2},{}^{18}O_{0;1}]CH_3-CH_2-OH$ $[2-{}^2H_{2;2},{}^{18}O_{0;1}]$Ethanol

R-8.3.3   **Nonselectively labelled compounds**

An isotopically labelled compound is designated as *nonselectively labelled* when the position(s) and the number of the labelling nuclide(s) are both undefined.

Note: When only atoms of an element to be modified are present at the same position in a compound, only specific or selective labelling can result. Nonselective labelling requires that the element to be modified be at different positions in the structure. For example, CH_4 and $CCl_3-CH_2-CCl_3$ can only be specifically or selectively labelled with a hydrogen isotope (see R-8.3.1.3 and R.8.3.2.1).

R-8.3.3.1   Nonselective labelling is indicated in the **formula** by inserting the nuclide symbol, enclosed in square brackets, directly before the usual line formula with no locants or subscripts.

Example:
$[{}^{13}C]CH_3-CH_2-CH_2-COOH$

R-8.3.3.2   The **name** of a nonselectively labelled compound is formed in the same way as the name of a selectively labelled compound but contains neither locants nor subscripts in the nuclide descriptor.

Examples:
Chloro$[{}^3H]$benzene
$[{}^{14}C]$Glycerol

R-8.3.4   **Isotopically deficient compounds**

An isotopically labelled compound may be designated as *isotopically deficient* when the isotopic content of one or more elements has been depleted, i.e., when one or more nuclide(s) is(are) present in less than the natural ratio.

R-8.3.4.1. Isotopic deficiency is denoted in the ***formula*** by adding the italicized syllable '*def*' immediately preceding, without a hyphen, the appropriate nuclide symbol.

Example:
[*def*^{13}C]CHCl$_3$

Note: According to one's viewpoint, one may also use [^{12}C]CHCl$_3$.

R-8.3.4.2 The ***name*** of an isotopically deficient compound may be formed by adding the italicized syllable *def* immediately preceding, without a hyphen, the appropriate nuclide symbol, both enclosed in square brackets and cited before the name or that part of the name that is isotopically modified.

Example:
[*def*^{13}C]Chloroform

Table 18. Comparative examples of formulae and names for isotopically modified compounds

Type of Compound	Formula	Name
Unmodified	CH$_3$–CH$_2$–OH	Ethanol
Isotopically substituted	C^{2}H$_3$–CH$_2$–O^{2}H	(2,2,2-^{2}H$_3$)Ethan(^{2}H)ol (*O*,2,2,2-^{2}H$_4$)Ethanol
Specifically labelled	C[^{2}H$_3$]–CH$_2$–O[^{2}H]	[2,2,2-^{2}H$_3$]Ethan[^{2}H]ol [*O*,2,2,2-^{2}H$_4$]Ethanol
Selectively labelled	[*O*,2-^{2}H]CH$_3$–CH$_2$–OH [2-^{2}H$_{2;2}$,^{18}O$_{0;1}$]CH$_3$–CH$_2$–OH	[*O*,2-^{2}H]Ethanol [2-^{2}H$_{2;2}$,^{18}O$_{0;1}$]Ethanol
Nonselectively labelled	[^{2}H]CH$_3$–CH$_2$–OH	[^{2}H]Ethanol
Isotopically deficient	[*def*^{13}C]CH$_3$–CH$_2$–OH	[*def*^{13}C]Ethanol

R-9   Appendix

R-9.0   INTRODUCTION

Trivial and semisystematic names for parent structures and parent substituent prefixes retained for use in naming organic compounds are given in R-9.1. Names used for describing bridges in naming bridged fused ring structures are given in R-9.2 and a list of replacement or 'a' prefixes is given in R-9.3.

R-9.1   TRIVIAL AND SEMISYSTEMATIC NAMES RETAINED FOR NAMING ORGANIC COMPOUNDS

The following tables list trivial and semisystematic names for parent structures and parent substituent prefixes[113] that are retained in these IUPAC recommendations for naming organic compounds. They are of three types:

● Type 1: Names that may be used when the structure is substituted at any position;

● Type 2: Names that may be used only when the structure is substituted in certain ways, for example, at a ring position;

● Type 3: Names that may not be used when the structure is substituted in any way.

The lists in the tables of R-9.1 along with names generated according to recommendations in R-2 are to be considered limiting unless specified otherwise; however, use of trivial and semisystematic names for compounds covered by special rules, for example, amino acids, carbohydrates, steroids, and, in general, natural products and related compounds, is allowed. Contracted forms and substituent group names not derived by a systematic transformation are given. However, names of derivatives of parent structures formed by systematic transformation of the parent name, such as the addition of suffixes, are not included. Not all the names for parent ring structures given in the tables may be used to generate names for fused-ring structures; a comprehensive document dealing with nomenclature for fused- and bridged-ring parent structures is in preparation.

Table 19 Acyclic and monocyclic hydrocarbons

(a) Parent Hydrocarbons

Type 1—Unlimited substitution

CH_4	CH_3-CH_3	$CH \equiv CH$
Methane	Ethane	Acetylene
$CH_3-CH_2-CH_3$	$CH_2=C=CH_2$	$CH_3-[CH_2]_2-CH_3$
Propane	Allene	Butane

Benzene

---

[113] '-Ylidene' and '-ylidyne' suffixes, where appropriate, are used in the same way as the '-yl' suffixes given.

Table 19 (Continued)

Type 2—Limited substitution (ring only)

C_6H_5–CH_3	C_6H_5–$CH=CH_2$	C_6H_5–$CH=CH$–C_6H_5
Toluene[a, b]	Styrene[a, b]	Stilbene[a, b]

Type 3—No substitution

$(CH_3)_2CH$–CH_3	$(CH_3)_2CH$–CH_2–CH_3	$(CH_3)_4C$
Isobutane	Isopentane	Neopentane

$CH_2=CH(CH_3)$–$CH=CH_2$
Isoprene

Fulvene[a]

$C_6H_4(CH_3)_2$
Xylene[a] (*o*-, *m*-, and *p*-isomers)

Mesitylene[a]

C_6H_5–$CH(CH_3)_2$
Cumene[a]

CH_3–C_6H_4–$CH(CH_3)_2$
Cymene[a] (*o*-, *m*- and *p*-isomers)

[a] In the 1979 edition of the IUPAC *Nomenclature of Organic Chemistry*[1], these names were used as parent hydrides with unlimited substitution.
[b] Only for substituents cited as prefixes.

(b) Substituent groups

Type 1—Unlimited substitution

–CH_2–	–CH_2–CH_2–	$CH_2=CH$–
Methylene[a]	Ethylene[b]	Vinyl
$CH_2=CH$–CH_2–	C_6H_5–	–C_6H_4–
Allyl	Phenyl	Phenylene

Type 2—Limited substitution (ring only)

C_6H_5–CH_2–	C_6H_5–$CH=$	C_6H_5–$CH=CH$–
Benzyl	Benzylidene[c]	Styryl[d]
C_6H_5–CH_2–CH_2–	C_6H_5–$CH=CH$–CH_2–	$(C_6H_5)_2CH$–
Phenethyl[d]	Cinnamyl	Benzhydryl
$(C_6H_5)_3C$–		
Trityl		

Type 3—No substitution

$(CH_3)_2CH$–	$(CH_3)_2C=$	$CH_2=C(CH_3)$–
Isopropyl	Isopropylidene[c]	Isopropenyl
$(CH_3)_2CH$–CH_2–	CH_3–CH_2–$CH(CH_3)$–	$(CH_3)_3C$–
Isobutyl	*sec*-Butyl	*tert*-Butyl
$(CH_3)_2CH$–$[CH_2]_2$–	CH_3–CH_2–$C(CH_3)_2$–	$(CH_3)_3C$–CH_2–
Isopentyl	*tert*-Pentyl	Neopentyl
CH_3–C_6H_4–	$2,4,6$–$(CH_3)_3C_6H_2$–	
Tolyl	Mesityl	

[a] The name methylene is used instead of methanediyl; the name methylidene is used for the $CH_2=$ group when doubly bonded to another part of the structure.
[b] The name ethylene is to be used for the bivalent substituent group CH_2–CH_2–, and not for the hydrocarbon $CH_2=CH_2$ which is named 'ethene' as in the 1979 edition of the IUPAC *Nomenclature of Organic Chemistry*[1].
[c] As an exception to R-2.5, these names are used in describing ketals and acetals derived from unsubstituted acetone and benzaldehyde, respectively, even though no double bond is present.
[d] In the 1979 edition of the IUPAC *Nomenclature of Organic Chemistry*[1], these names were used as parent hydrides with unlimited substitution.

Table 20 Unsaturated polycyclic hydrocarbons

Type 1—Unlimited substitution

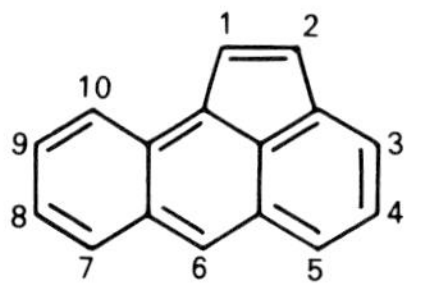

Aceanthrylene

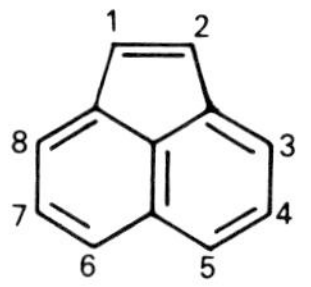

Acenaphthylene

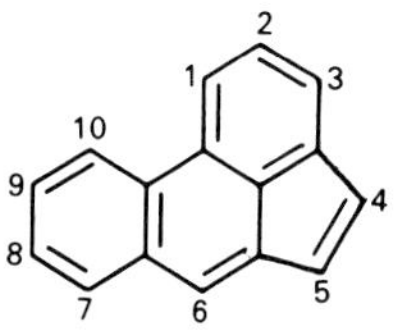

Acephenanthrylene

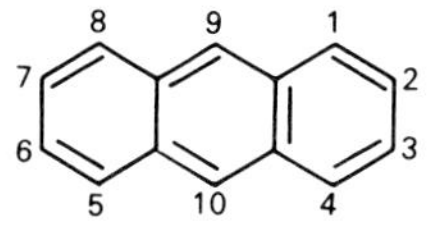

Anthracene
(An exception to systematic numbering)

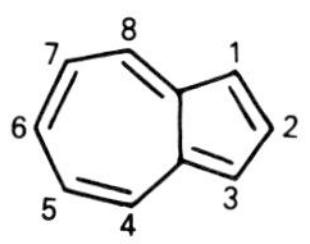

Azulene

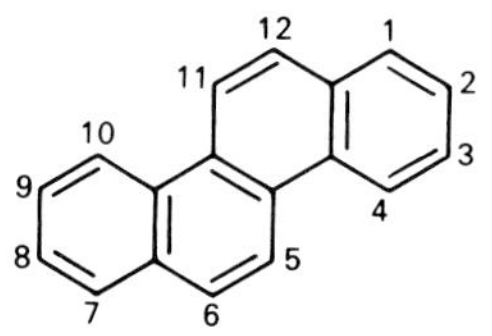

Chrysene

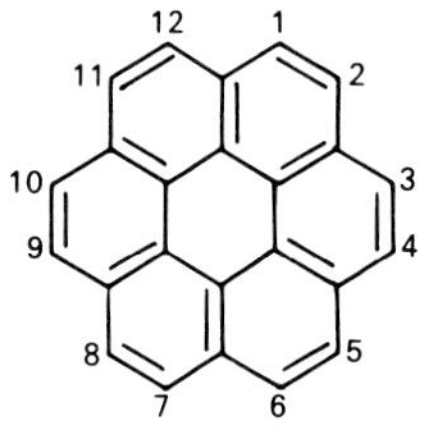

Coronene

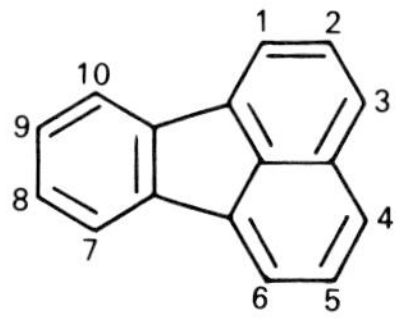

Fluoranthene

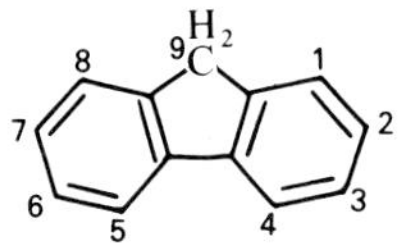

Fluorene (9*H*-isomer shown)[a]

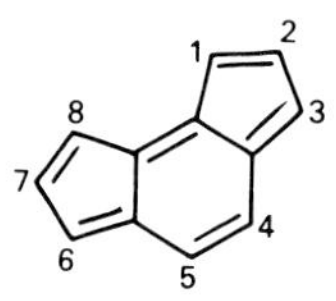

as-Indacene

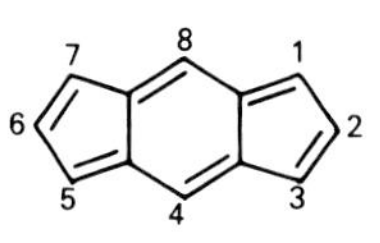

s-Indacene

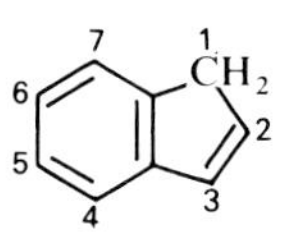

Indene (1*H*-isomer shown)[a]

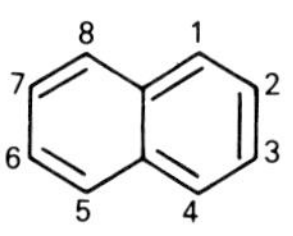

Naphthalene

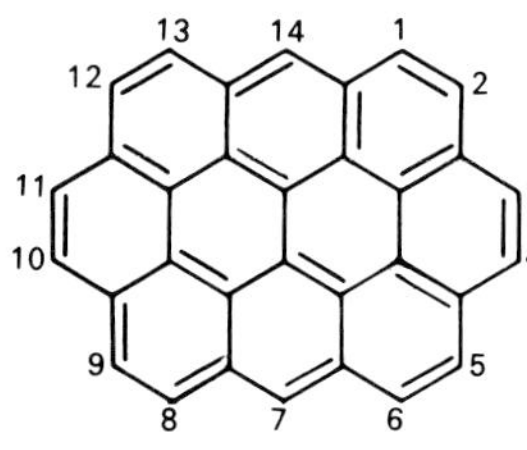

Ovalene

[a] The isomer shown is generally used without specifying the indicated hydrogen.

Table 20 (Continued)

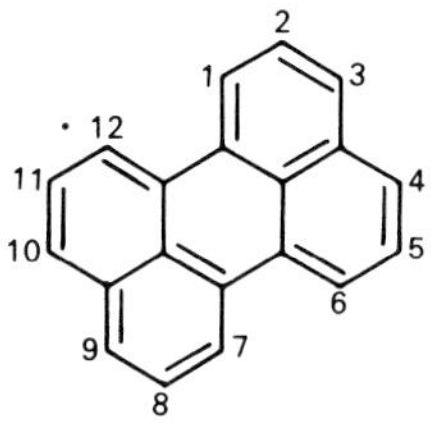

Perylene

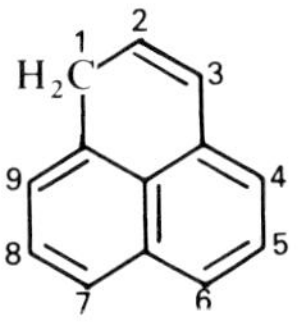

Phenalene (1*H*-isomer shown)[a]

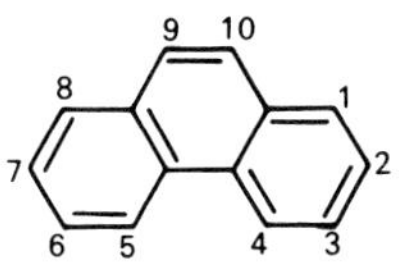

Phenanthrene
(An exception to systematic numbering)

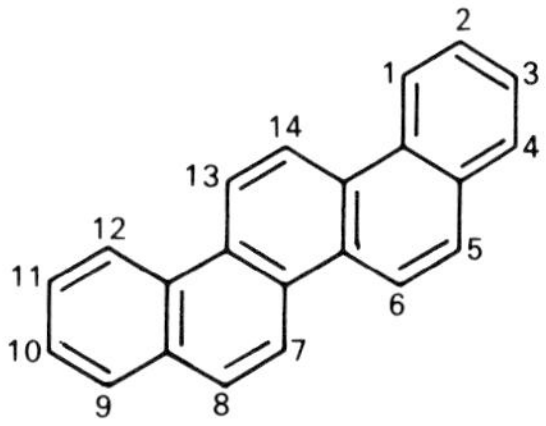

Picene

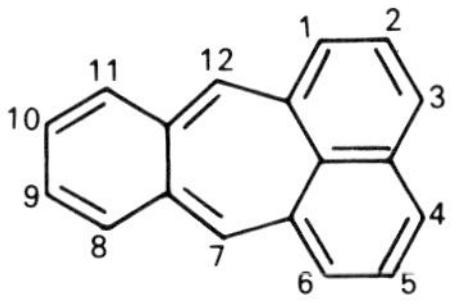

Pleiadene

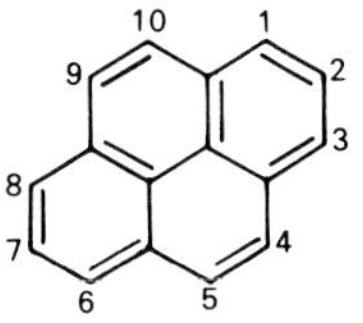

Pyrene

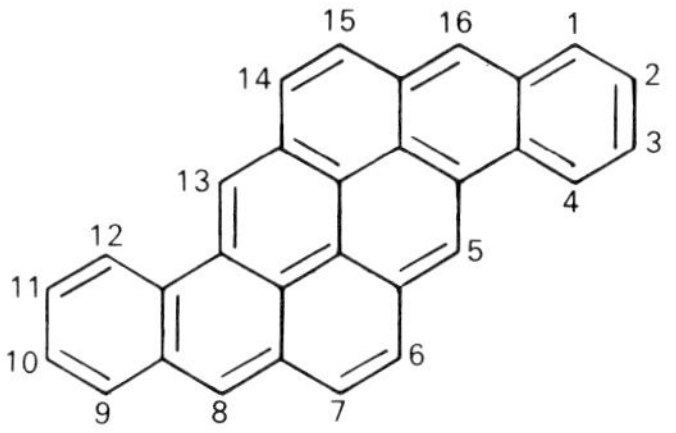

Pyranthrene

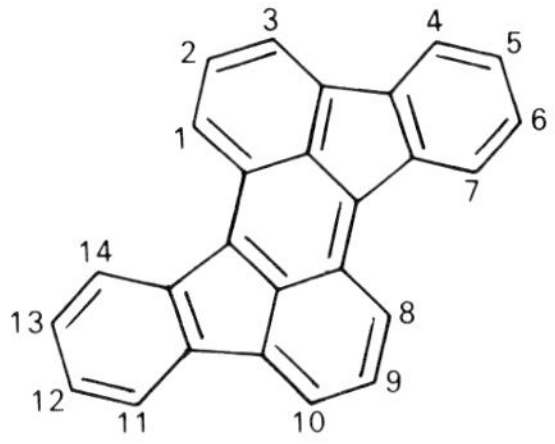

Rubicene

[a] The isomer shown is generally used without specifying the indicated hydrogen.

PRIORITY ORDER[114]

Pentalene	Fluorene	Chrysene	Hexacene
Indene	Phenalene	Tetracene[115]	Rubicene
Naphthalene	Phenanthrene	Pleiadene	Coronene
Azulene	Anthracene	Picene	Trinaphthylene
Heptalene	Fluoranthene	Perylene	Heptaphene
Biphenylene	Acephenanthrylene	Pentaphene	Heptacene
as-Indacene	Aceanthrylene	Pentacene	Pyranthrene
s-Indacene	Triphenylene	Tetraphenylene	Ovalene
Acenaphthylene	Pyrene	Hexaphene	

[114] An illustrative list in *increasing* order of precedence following each column downward in turn for choice as the principal component in fusion nomenclature (see Rule A-21.2, pp. 20–22 in IUPAC *Nomenclature of Organic Chemistry*[1]). Note that there are names in this list that do not appear in Table 20.
[115] Known as naphthacene in IUPAC *Nomenclature of Organic Chemistry*[1]

Table 21 Saturated polycyclic hydrocarbons

Type 1—Unlimited substitution

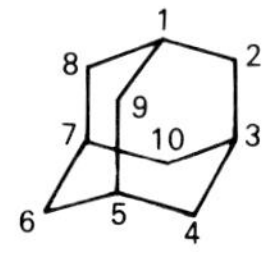

Adamantane

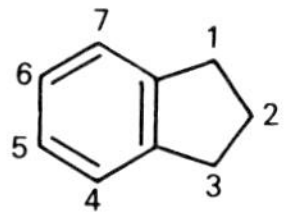

Indane (formerly Indan, see ref. 1)

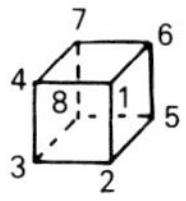

Cubane

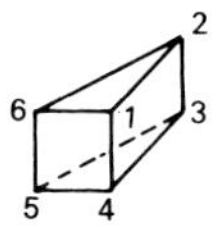

Prismane

Table 22 Polycyclic hydrocarbon substituent prefixes

Type 1—Unlimited substitution

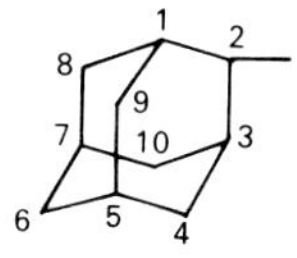

Adamantyl (2-isomer shown)

Anthryl (2-isomer shown)
(An exception to systematic numbering)

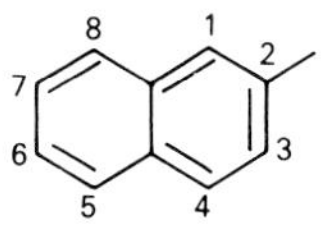

Naphthyl (2-isomer shown)

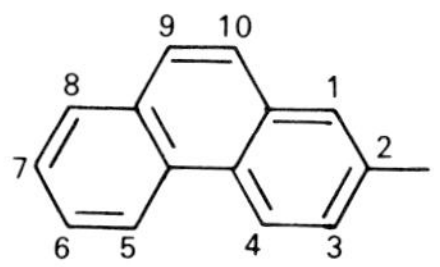

Phenanthryl (2-isomer shown)
(An exception to systematic numbering)

Table 23 Heterocyclic parent hydrides

Type 1—Unlimited substitution

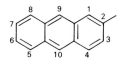

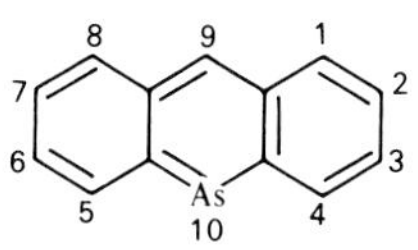

Acridarsine
(An exception to systematic numbering)

Acridine
(An exception to systematic numbering)

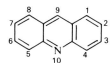

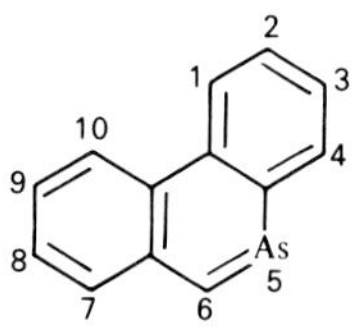

Arsanthridine

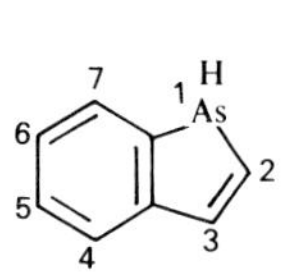

Arsindole (1*H*-isomer shown)

166

Table 23 (Continued)

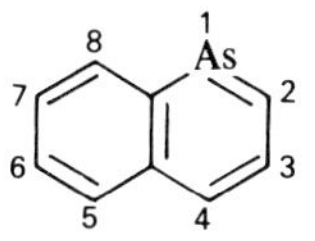

Arsinoline

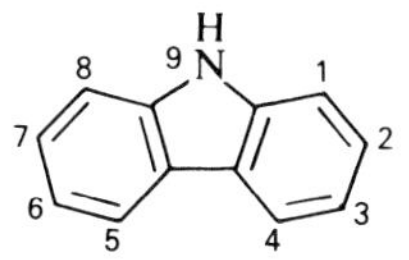

Carbazole (9H-isomer shown)
(An exception to systematic numbering)

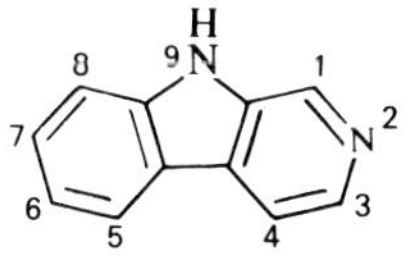

β-Carboline[116] (9H-isomer shown)

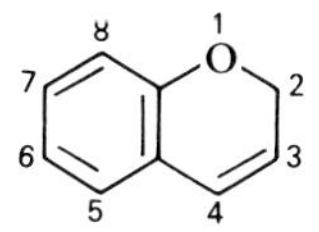

Chromene (2H-isomer shown)
(Chalcogen analogues are named using
the prefixes 'thio-', 'seleno-' and 'telluro-'.)

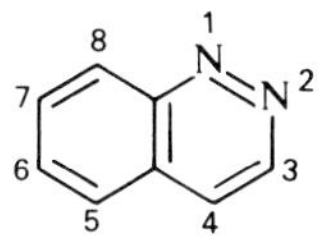

Cinnoline

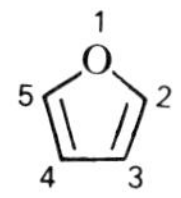

Furan

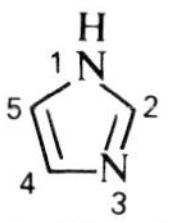

Imidazole (1H-isomer shown)[a]

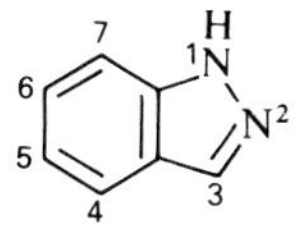

Indazole (1H-isomer shown)[a]

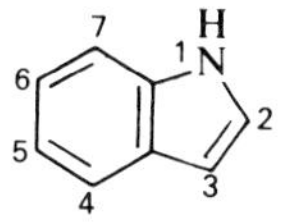

Indole (1H-isomer shown)[a]

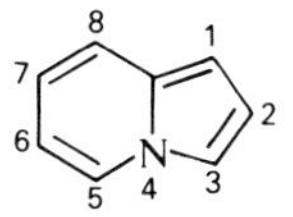

Indolizine

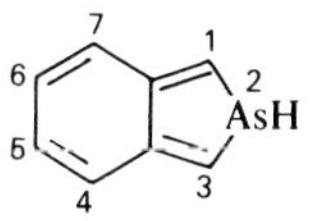

Isoarsindole (2H-isomer shown)

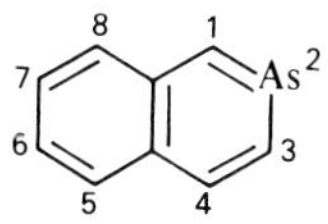

Isoarsinoline

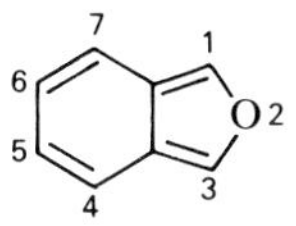

Isobenzofuran

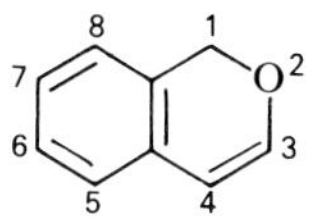

Isochromene[117] (1H-isomer shown)[a]
(Chalcogen analogues are named using
the prefixes 'thio-', 'seleno-' and 'telluro-').

[a] The isomer shown is generally used without specifying the indicated hydrogen.

[116] Although, according to the introduction to Table IV in the Appendix to Section D of IUPAC *Nomenclature of Organic Chemistry*[1] (see p. 466), this name was to be deleted from the list of trivial and semisystematic names given in B-2.11, pp. 55–61, it is included in these recommendations.
[117] This trivial name was not included in the 1979 edition of IUPAC *Nomenclature of Organic Chemistry*[1].

Table 23 (Continued)

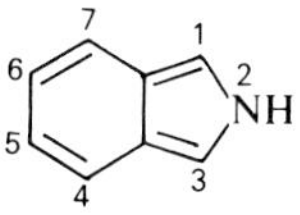

Isoindole (2*H*-isomer shown)[a]

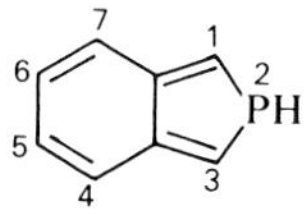

Isophosphindole (2*H*-isomer shown)

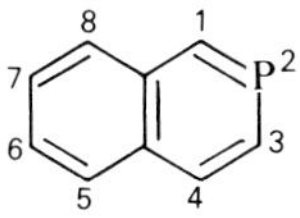

Isophosphinoline

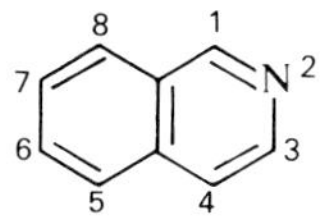

Isoquinoline

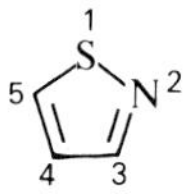

Isothiazole

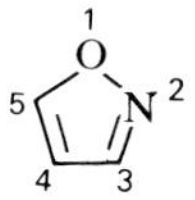

Isoxazole

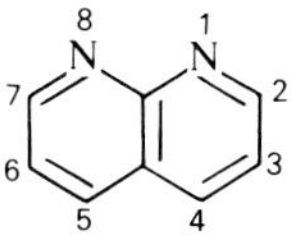

Naphthyridine (1,8-isomer shown)

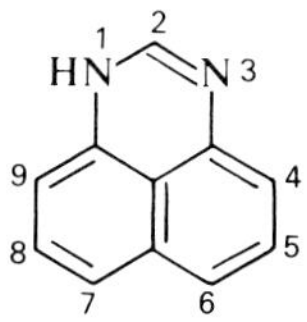

Perimidine (1*H*-isomer shown)[a]

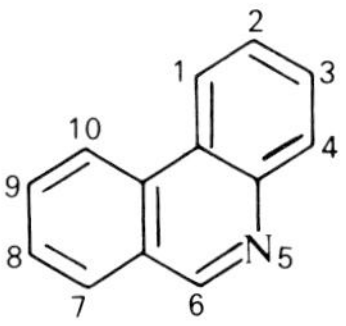

Phenanthridine

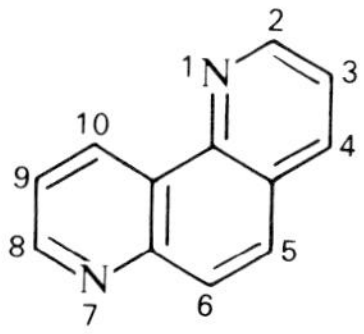

Phenanthroline (1,7-isomer shown)

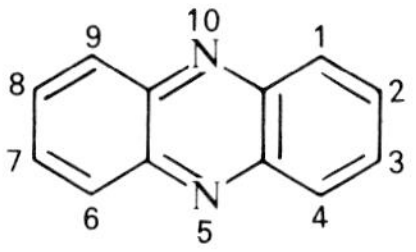

Phenazine

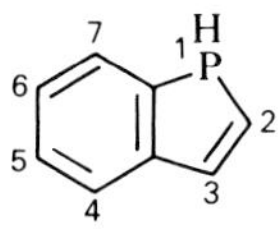

Phosphindole (1*H*-isomer shown)

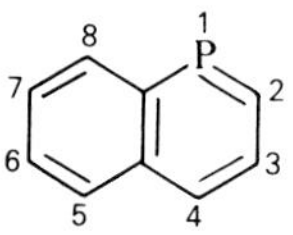

Phosphinoline

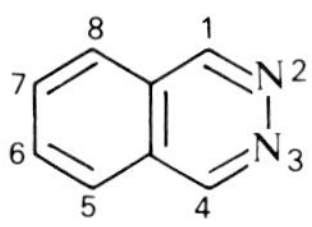

Phthalazine

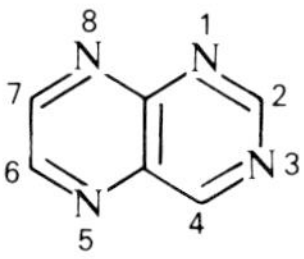

Pteridine

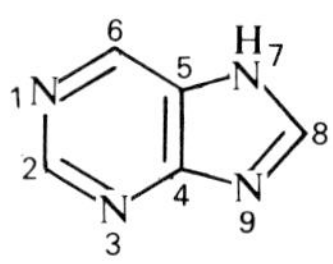

Purine (7*H*-isomer shown)[a]
(An exception to systematic numbering)

[a] The isomer shown is generally used without specifying the indicated hydrogen.

Table 23 (Continued)

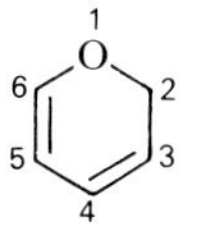

Pyran (2H-isomer shown)
(Chalcogen analogues are named using
the prefixes 'thio-', 'seleno-', and 'telluro-'.)

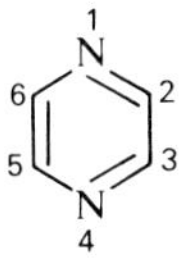

Pyrazine

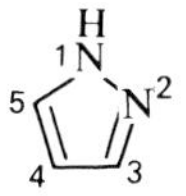

Pyrazole (1H-isomer shown)[a]

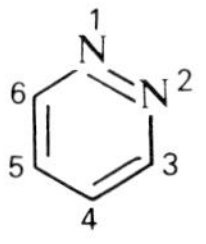

Pyridazine

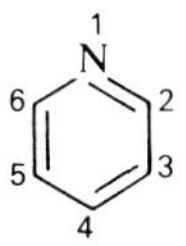

Pyridine

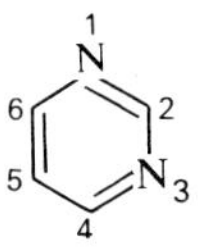

Pyrimidine

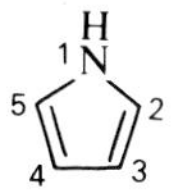

Pyrrole (1H-isomer shown)[a]

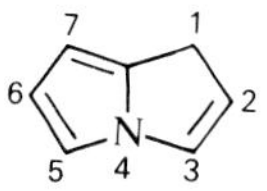

Pyrrolizine (1H-isomer shown)

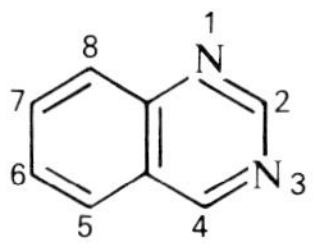

Quinazoline

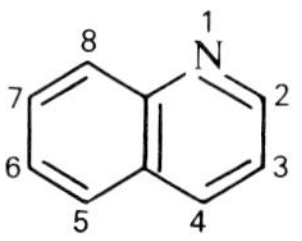

Quinoline

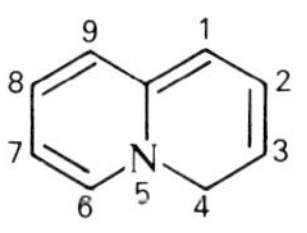

Quinolizine (4H-isomer shown)

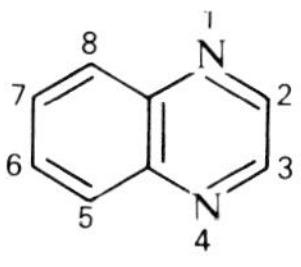

Quinoxaline

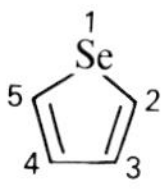

Selenophene

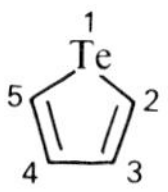

Tellurophene

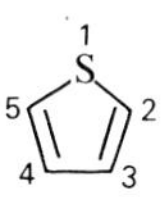

Thiophene

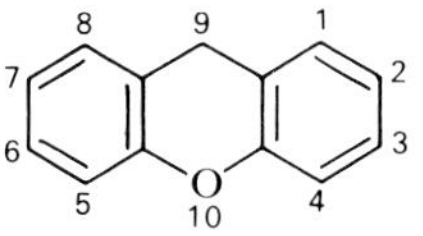

Xanthene (9H-isomer shown)[a]
(An exception to systematic numbering)
(Chalcogen analogues are named using
the prefixes 'thio-', 'seleno-' and 'telluro-'.)

[a] The isomer shown is generally used without specifying the indicated hydrogen.

PRIORITY OF PRECEDENCE[118,119]

Phenomercurine	Phenothiarsine	Isoxazole	Quinazoline
Isoarsindole	Furan	Pyridine	Cinnoline
Arsindole	Pyran	Pyrazine	Pteridine
Isoarsinoline	Isobenzofuran	Pyrimidine	Carbazole
Arsinoline	Isochromene	Pyridazine	β-Carboline
Arsanthridine	Chromene	Pyrrolizine	Phenanthridine
Acridarsine	Xanthene	Indolizine	Acridine
Arsanthrene	Phenoxantimonine	Isoindole	Perimidine
Isophosphindole	Phenoxarsine	Indole	Phenanthroline
Phosphindole	Phenoxaphosphine	Indazole	Phenazine
Isophosphinoline	Phenoxatellurine	Purine	Phenomercazine
Phosphinoline	Phenoxaselenine	Quinolizine	Phenarsazine
Phosphanthrene	Phenoxathiine	Isoquinoline	Phenophosphazine
Tellurophene	Pyrrole	Quinoline	Phenotellurazine
Selenophene	Imidazole	Phthalazine	Phenoselenazine
Selenanthrene	Pyrazole	Naphthyridine	Phenothiazine
Thiophene	Isothiazole	Quinoxaline	Phenoxazine
Thianthrene			

[118] An illustrative list in *increasing* order of precedence following each column downward in turn for choice as the principal component in fusion nomenclature (see Rule B-2.11, pp. 55–61 and Table IV in the Appendix to Section D, pp. 466–471 in IUPAC *Nomenclature of Organic Chemistry*[1]). Note that there are names in this list that do not appear in Table 23.

[119] Since the final 'e' of Hantzsch–Widman names, on which these names are based, is optional (see Note 2 to Table 4, p. 42), the final 'e' of the names in this list based on Hantzsch–Widman names is also optional.

Table 24 Hydrogenated heterocyclic parent hydrides[120]

Type 1—Unlimited substitution

Chromane
(Chalcogen analogues are named using the prefixes 'thio-', 'seleno-', and 'telluro-'.)

Imidazolidine

Indoline

Isochromane
(Chalcogen analogues are named using the prefixes 'thio-', 'seleno-', and 'telluro-'.)

Isoindoline

Morpholine
(Chalcogen analogues are named using the prefixes 'thio-', 'seleno-', and 'telluro-').

Piperazine

Piperidine

Pyrazolidine

Pyrrolidine

Quinuclidine

---

[120] Pyrroline, imidazoline, and pyrazoline, which were in previous editions of the IUPAC *Nomenclature of Organic Chemistry*[1], are not included in these recommendations. They are named on the basis of fully unsaturated Hantzsch–Widman names (see R-2.3.3.3).

Table 25 Heterocyclic substituent groups and cations

Type 1—Unlimited substitution

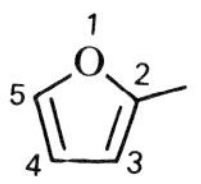

Furyl (2-isomer shown)

Isoquinolyl (3-isomer shown)

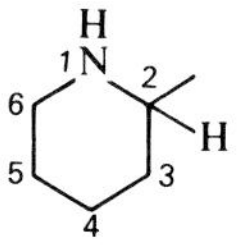

Piperidyl (2-isomer shown)

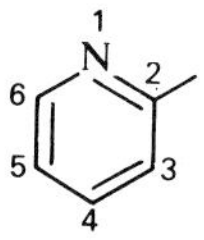

Pyridyl (2-isomer shown)

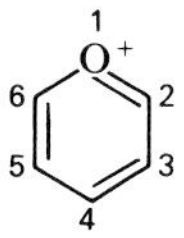

Pyrylium

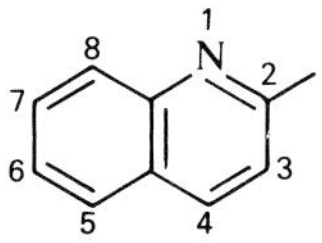

Quinolyl (2-isomer shown)

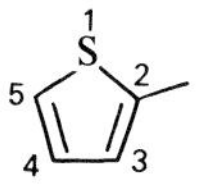

Thienyl (2-isomer shown)

Type 2—Limited substitution (ring only)

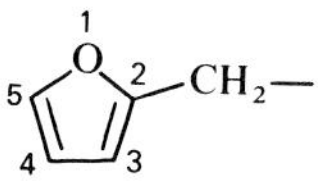

Furfuryl (2-isomer *only*)

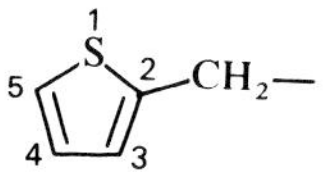

Thenyl (2-isomer *only*)

Table 26 Hydroxy compounds, ethers, and related substituent groups

(a) Parent structures

Type 1—Unlimited substitution
C_6H_5–OH
Phenol

Type 2—Limited substitution (ring only)
C_6H_5–O–CH_3
Anisole

Type 3—No substitution[121]

HO–CH_2–CH_2–OH
Ethylene glycol

$C(CH_2OH)_4$
Pentaerythritol

HO–CH_2–CH(OH)–CH_2–OH
Glycerol

$(CH_3)_2C(OH)$ –C(OH) $(CH_3)_2$
Pinacol[122]

CH_3—⟨ring⟩—OH

Cresol (*p*-isomer shown)

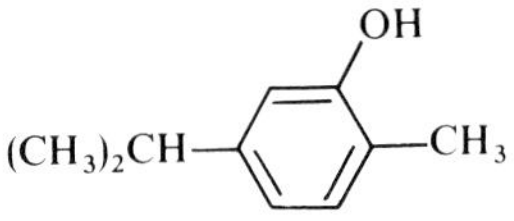

Carvacrol

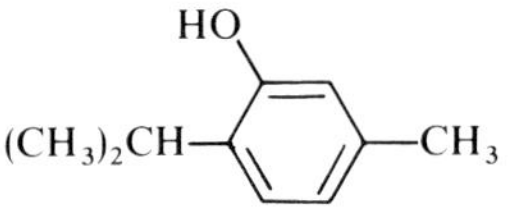

Thymol

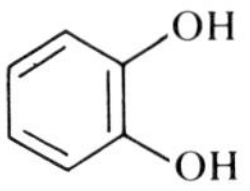

Pyrocatechol

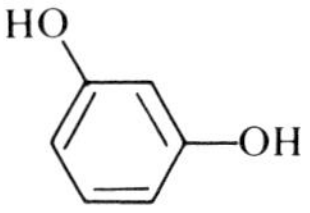

Resorcinol

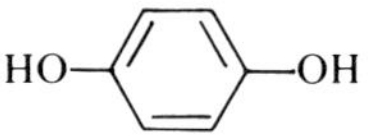

Hydroquinone

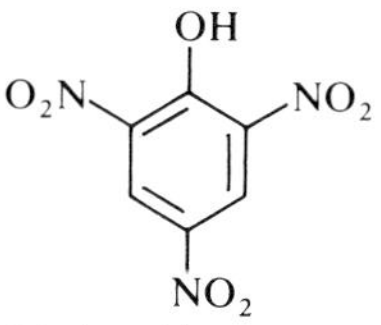

Picric acid

(b) Substituent groups

Type 1—Unlimited substitution
CH_3–O–
Methoxy

CH_3–CH_2–O–
Ethoxy

CH_3–$[CH_2]_2$–O–
Propoxy

CH_3–$[CH_2]_3$–O–
Butoxy

C_6H_5–O–
Phenoxy

Type 3—No substitution
$(CH_3)_2CH$–O–
Isopropoxy

$(CH_3)_2CH$–CH_2–O–
Isobutoxy

CH_3–CH_2–$CH(CH_3)$–O–
sec-Butoxy

$(CH_3)_3C$–O–
tert-Butoxy

---

[121] Replacement of the hydrogen atom of a hydroxy group may be considered to be a functionalization rather than a substitution, for instance, in the formation of an ester, and is allowed.
[122] The name 'pinacol' has also been used as a class name.

Table 27 Carbonyl compounds and derived substituent groups[a]

(a) Parent structures

Type 1—Unlimited substitution

$CH_3–CO–CH_3$
Acetone

$CH_2=C=O$
Ketene

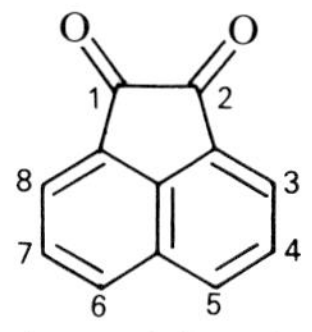

Acenaphthoquinone (1,2-isomer *only*)

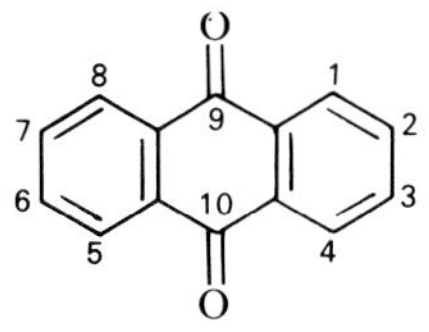

Anthraquinone (9,10-isomer shown)

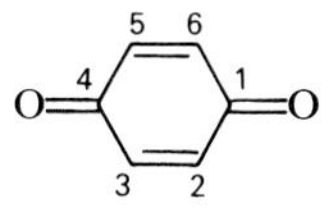

Benzoquinone (*p*-isomer shown)

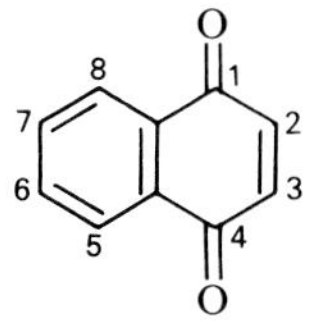

Naphthoquinone (1,4-isomer shown)

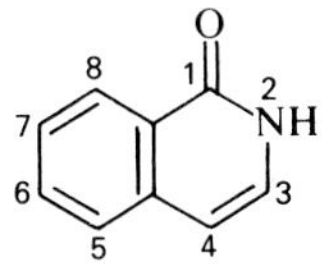

Isoquinolone (1-isomer shown)
(Abbreviated form of the systematic
'-inone' ending)

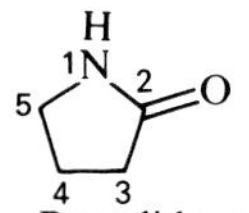

Pyrrolidone (2-isomer shown)
(Abbreviated form of the systematic
'-idinone' ending)

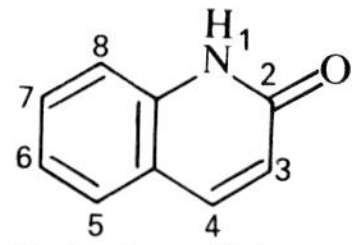

Quinolone (2-isomer shown)
(Abbreviated form of the systematic
'-inone' ending)

Type 3—No substitution

$C_6H_5–CO–CH_3$
Acetophenone[b]

$C_6H_5–CO–C_6H_5$
Benzophenone[b]

$C_6H_5–CH=CH–CO–C_6H_5$
Chalcone

$C_6H_5–CO–CO–C_6H_5$
Benzil[b]

$CH_3–CO–CO–CH_3$
Biacetyl[b]

$C_6H_5–CO–CH_2–CH_3$
Propiophenone[b]

(b) Substituent groups

Type 1—Unlimited Substitution
$CH_3–CO–CH_2–$
Acetonyl

Type 3—No Substitution
$C_6H_5–CO–CH_2–$
Phenacyl

[a] For aldehydes, see footnote 123 (3).
[b] In previous editions of the IUPAC *Nomenclature of Organic Chemistry*[1], unlimited substitution was permitted on these ketone names.

Table 28 Carboxylic acids and related groups[123]

(a) Unsubstituted parent structures

Type 1—Unlimited substitution

CH_3–COOH
Acetic acid†

CH_2=CH–COOH
Acrylic acid

HOOC–CH_2–COOH
Malonic acid†

HOOC–$[CH_2]_2$–COOH
Succinic acid†

$$HC-COOH$$
$$\|$$
$$HOOC-CH$$

C_6H_5–COOH
Benzoic acid

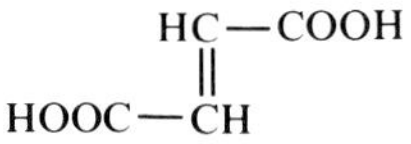

Fumaric acid

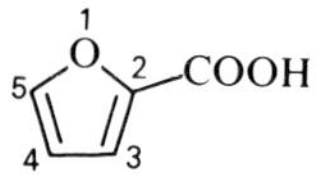

Furoic acid (2-isomer shown)

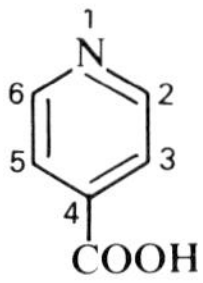

Isonicotinic acid

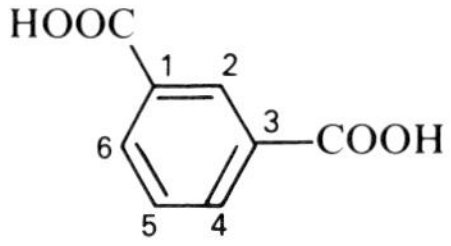

Isophthalic acid

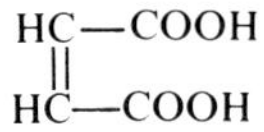

$$HC-COOH$$
$$\|$$
$$HC-COOH$$

Maleic acid

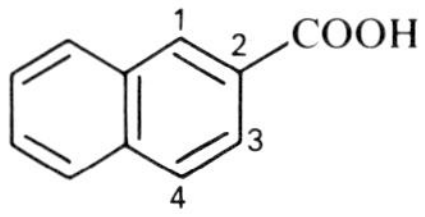

Naphthoic acid (2-isomer shown)

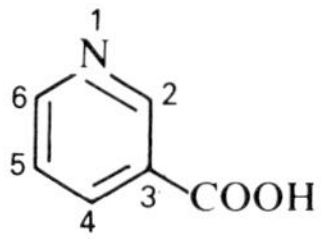

Nicotinic acid

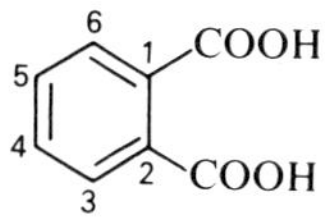

Phthalic acid

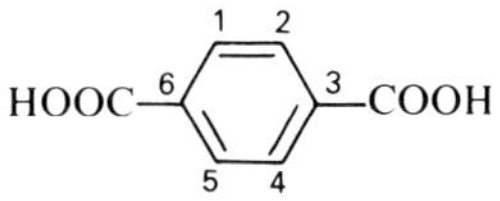

Terephthalic acid

Type 3—No substitution[124]

H–COOH
Formic acid†

CH_3–CH_2–COOH
Propionic acid†

CH≡C–COOH
Propiolic acid

CH_3–$[CH_2]_2$–COOH
Butyric acid†

---

† For meaning of the 'dagger' (†), see footnote 123 (1).

[123] Related groups include acyl substituent groups, acid halides, anhydrides, hydrazides, amides, imides, nitriles, and aldehydes whose names are formed from the appropriate name stem as follows:

(1) *acyl groups* by changing the '-ic acid' or '-oic acid' ending to '-oyl' (except for those names in the table marked with a dagger (†), for which the '-ic acid' ending is changed to '-yl') (see also R-5.7.1).

(2) *acid halides* by adding the appropriate class name to the name of the acyl group (see also R-5.7.6).

(3) *aldehydes and amides* by replacing the '-ic acid' or '-oic acid' ending by '-aldehyde' or 'amide', respectively (see also R-5.6.1 or R-5.7.8.1, respectively).

(4) *nitriles and hydrazides* by replacing the '-ic acid' or '-oic acid' ending by '-onitrile' or '-ohydrazide', respectively (see also R-5.7.9 or R-5.7.8.4, respectively).

(5) *imides* by replacing the '-ic acid' ending of a dicarboxylic acid by '-imide' (see also R-5.7.8.3).

[124] Replacement of the hydrogen atom of a carboxy group may be considered to be functionalization rather than substitution, for instance, in the formation of a salt, ester, or anhydride, and is allowed.

Table 28 (Continued)

(CH₃)₂CH–COOH	CH₂=C(CH₃)–COOH

$(CH_3)_2CH-COOH$
Isobutyric acid†

$CH_2=C(CH_3)-COOH$
Methacrylic acid

$CH_3-[CH_2]_{14}-COOH$
Palmitic acid

$CH_3-[CH_2]_{16}-COOH$
Stearic acid

$CH_3-[CH_2]_7-CH=CH-[CH_2]_7-COOH$
Oleic acid

$HOOC-COOH$
Oxalic acid[125,]†

$HOOC-[CH_2]_3-COOH$
Glutaric acid†

$HOOC-[CH_2]_4-COOH$
Adipic acid

$C_6H_5-CH=CH-COOH$
Cinnamic acid

(b) Hydroxy, oxo, and amino (not α-amino) carboxylic acids

Type 3—No substitution[124]
$HO-CH_2-COOH$
Glycolic acid

$CH_3-CH(OH)-COOH$
Lactic acid

$HO-CH_2-CH(OH)-COOH$
Glyceric acid

$HOOC-[CH(OH)]_2-COOH$
Tartaric acid

$$\begin{array}{c} CH_2-COOH \\ | \\ HO-C-COOH \\ | \\ CH_2-COOH \end{array}$$
Citric acid

$OHC-COOH$
Glyoxylic acid

$CH_3-CO-COOH$
Pyruvic acid

$CH_3-CO-CH_2-COOH$
Acetoacetic acid

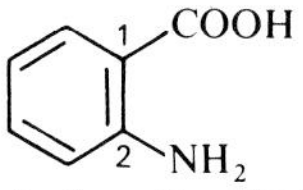

Anthranilic acid (1,2-isomer *only*)

$(C_6H_5)_2C(OH)-COOH$
Benzilic acid

$(HOOC-CH_2)_2N-[CH_2]_2-N(CH_2-COOH)_2$
Ethylenediaminetetraacetic acid

(c) Amic acids and peroxy carboxylic acids

Type 1—Unlimited substitution
$H_2N-COOH$
Carbamic acid

$H_2N-CO-COOH$
Oxamic acid

Type 3—No substitution[124]
$HCO-OOH$
Performic acid

$CH_3-CO-OOH$
Peracetic acid

$C_6H_5-CO-OOH$
Perbenzoic acid

---

† For meaning of the 'dagger' (†) see footnote 123 (1).

[125] The corresponding amide is oxamide (a contracted form) and the corresponding aldehyde is glyoxal.

Table 29 Amines, nitrogenous heterocyclic parent structures, and derived substituent groups

(a) Parent structures

Type 1—Unlimited substitution

$C_6H_5-NH_2$
Aniline

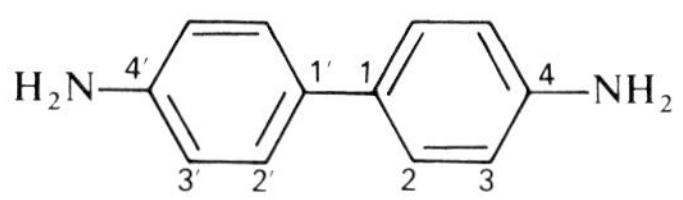

Hydantoin

Rhodanine

H₂N—(biphenyl)—NH₂
Benzidine (4,4'-isomer *only*)

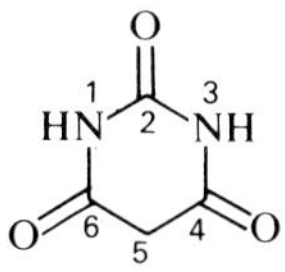

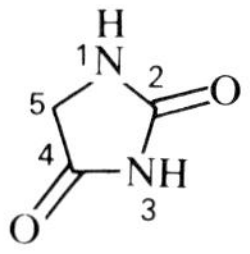

Barbituric acid

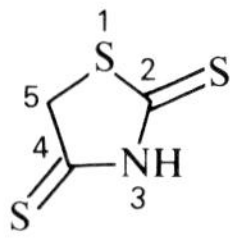

Type 3—No substitution

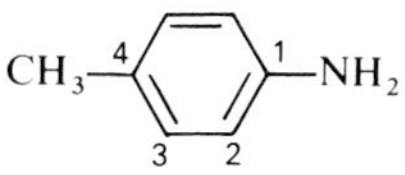

CH_3—(ring)—NH_2
Toluidine (*p*-isomer shown)

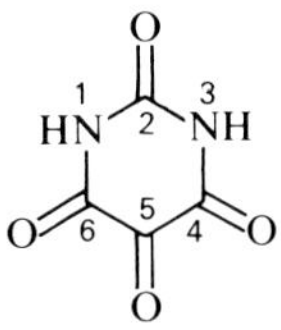

Alloxane

(b) Derived substituent groups

Type 1—Unlimited substitution

C_6H_5-NH-
Anilino

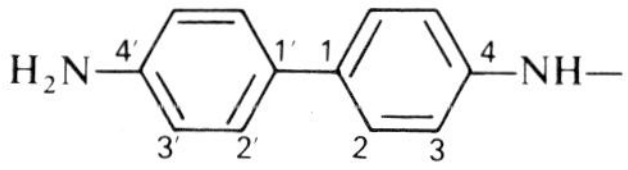

H₂N—(biphenyl)—NH—
Benzidino (4,4'-isomer *only*)

Type 3—No substitution

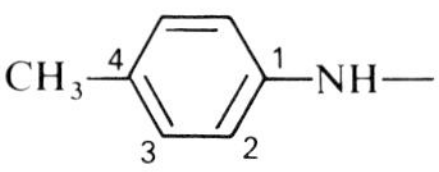

CH_3—(ring)—$NH-$
Toluidino (*p*-isomer shown)

Table 30 Sulfides, sulfonic acids, and derived substituent groups

(a) Parent structures

Type 1—Unlimited substitution

$(H_2N–CS)_2S$
Thiuram monosulfide

$(H_2N–CS)_2S_2$
Thiuram disulfide

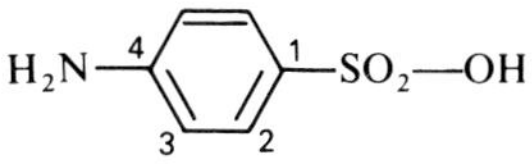

Sulfanilic acid (*p*-isomer *only*)

(b) Derived substituent groups

Type 3—No substitution

$CH_3–SO_2–$
Mesyl[126]

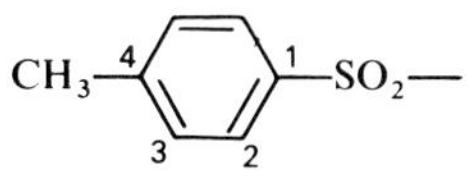

Tosyl[126] (*p*-isomer *only*)

---

[126] These names are abbreviations derived from names of the corresponding acyl groups, for example, methanesulfonyl, and are not to be used to create the name of the acid.

Table 31 Acyclic polynitrogen parent structures and derived substituent groups

(a) Parent structures

Type 1—Unlimited substitution

$H_2N–CO–NH–COOH$
4 3 2 1

Allophanic acid[a]

$H_2N–CO–NH–CO–NH_2$
5 4 3 2 1

Biuret[a]

$HN=N–CO–N=NH$
5 4 3 2 1

Carbodiazone[a]

$H_2N–NH–CO–NH–NH_2$
5 4 3 2 1

Carbonohydrazide[a,b]

$H_2N–C(=NH)–NH_2$
3 2 1

Guanidine[a]

$NH_2–NH_2$
2 1

Hydrazine

$H_2N–[C(=NH)–NH]_n–C(=NH)–NH_2$
2n+3   2 1

    n = 2, 3....

Triguanide[a], Tetraguanide[a], etc.

$H_2N–CO–NH–NH_2$
4 3 2 1

Semicarbazide[a]

$H_2N–C(=NH)–NH–C(=NH)–NH_2$
5 4 3 2 1

Biguanide[a]

$HN=N–CO–NH–NH_2$
5 4 3 2 1

Carbazone[a]

$HN=C=NH$

Carbodiimide

$HN=N–CH=N–NH_2$
5 4 3 2 1

Formazan[a]

$H_2N–CO–NH–CH_2–COOH$
5 4 3 2 1

Hydantoic acid[a]

$HN=C(OH)–NH_2$
3 2 1

Isourea[a]

$H_2N–[CO–NH]_n–CO–NH_2$
2n+3   2 1

    n = 2, 3....

Triuret[a], Tetrauret[a], etc.

$H_2N–CO–NH_2$
3 2 1

Urea[a]

(b) Derived substituent groups

Type 1—Unlimited substitution

H_2N CO NH CO
4 3 2 1

Allophanyl[a]

$H_2N–C(=NH)–NH–$
3 2 1

Guanidino[a]

$NH_2–NH–$
2 1

Hydrazino

$H_2N–CO–NH–$
3 2 1

Ureido[a]

$HN=N–CO–NH–NH–$
5 4 3 2 1

Carbazono[a]

$H_2N–CO–NH–CH_2–CO–$
5 4 3 2 1

Hydantoyl[a]

$H_2N–CO–NH–NH–$
4 3 2 1

Semicarbazido[a]

$–NH–CO–NH–$
3 2 1

Ureylene[a]

[a] Special numbering.
[b] Preferred to either of the abbreviated names carbohydrazide or carbazide.

179

Table 32 Halogen compounds

Type 3—No substitution	
Fluoroform	CHF_3
Chloroform	$CHCl_3$
Bromoform	$CHBr_3$
Iodoform	CHI_3
Phosgene	$COCl_2$
Thiophosgene	$CSCl_2$[a]

[a] And similarly for other chalcogen analogues.

R-9.2   **BRIDGE NAMES**

R-9.2.1   **Simple bivalent bridges**

R-9.2.1.1   ***An acyclic bridge*** is named as a prefix derived from the hydrocarbon name by changing the final 'e' to 'o'. The locant of a double bond, if present, is indicated in square brackets between the hydrocarbon prefix and the ending '-eno', '-dieno', etc.

Examples:

$-CH_2-$	Methano
$-CH_2-CH_2-$	Ethano
$-CH_2-CH_2-CH_2-$	Propano
$-CH_2-CH_2-CH_2-CH_2-$	Butano
$-CH=CH-$	Etheno
$-CH=CH-CH_2-$ $\quad 1 \quad 2 \quad 3$	Prop[1]eno
$-CH=CH-CH_2-CH_2-$ $\quad 1 \quad 2 \quad 3 \quad 4$	But[1]eno
$-CH_2-CH=CH-CH_2-$ $\quad 1 \quad 2 \quad 3 \quad 4$	But[2]eno
$-CH=CH-CH=CH-$ $\quad 1 \quad 2 \quad 3 \quad 4$	Buta[1,3]dieno

R-9.2.1.2   ***A monocyclic hydrocarbon bridge*** other than *benzene* is named by the same prefix as used as a fusion prefix (see R-2.4.1) preceded by 'epi-'.[127] The bridge is assumed to have the maximum number of noncumulative double bonds consistent with the attachment to the fused ring and/or to other bridges. The positions of the free valences are indicated by the relevant locants in square brackets directly in front of the bridge name.

Examples:

[1,3]Epicyclopropa   [1,2]Epicyclopenta

---

[127] This is an extension of the use of 'epi–' in prefixes such as 'epithio–'. The prefix *'endo-'* has been used for the same purpose elsewhere (see IUPAC 'Nomenclature of Fused and Bridged Fused Ring Systems', *Pure and Appl. Chem.*, in preparation).

R-9.2.1.3   **Other cyclic hydrocarbon bridges** are named by prefixes derived from the unsaturated hydrocarbon name (see Table 20, p. 164). Locants enclosed in square brackets are used as needed.

Examples:

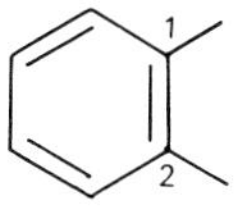

[1,2]Benzeno

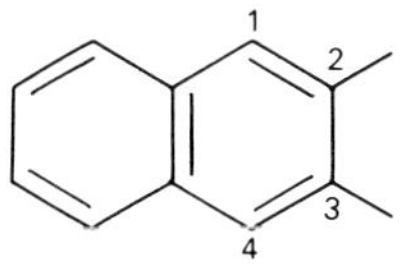

[2,3]Naphthaleno

R-9.2.1.4   **An acyclic heteroatomic bridge** is named by the appropriate prefix[128].

Examples:

–O–	Epoxy
–S–	Epithio
–Se–	Episeleno
–O–O–	Epidioxy
–S–S–	Epidithio
–SH$_2$–	λ^4-Sulfano
–O–S–	Epoxythio
–O–S–O–	Epoxythioxy
–O–NH–	Epoxyimino
–NH–	Epimino
–NH–NH–	Diazano[129]
–N=N–	Diazeno[129]
–N=N–NH–	Triaz[1]eno[129]
–PH–	Phosphano
–SnH$_2$–	Stannano
–O–CH$_2$–	Epoxymethano
–O–CH$_2$–CH$_2$–	Epoxyethano
–O–CH=CH–CH$_2$–	Epoxyprop[1]eno

R-9.2.1.5   **Heterocyclic bridges** are named by prefixes derived from the corresponding heterocyclic compound name (see Table 23, p. 166). Locants enclosed in square brackets are used where necessary.

Example:

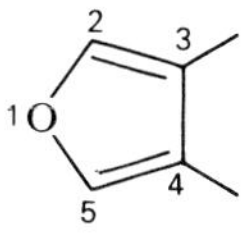

[3,4]Furano

---

[128] These bridge prefix names do not necessarily correspond to the names of the same divalent groups in other contexts.

[129] In the previous edition of the IUPAC *Nomenclature of Organic Chemistry*[1], these bridge prefixes were named 'biimino', 'azo', and 'azimino', respectively.

R-9.2.2   **Simple polyvalent bridges.** Tri- and tetravalent bridges derived from methane are named metheno and methyno, respectively. Other polyvalent bridges, monatomic or polyatomic, are named in the same way as the corresponding substituent prefix. Locants enclosed in square brackets are used as needed.

Examples:

–CH= or –CH–
Metheno[130]

–N= or –N–
Nitrilo[131]

–CH$_2$–CH=
Ethanylylidene

–CH$_2$–CH–
Ethane[1,1,2]triyl

–O–N= or –O–N–
Epoxynitrilo[131]

R-9.3   **'a' PREFIXES USED IN REPLACEMENT NOMENCLATURE**
These are listed below in decreasing order of priority following each column downward in turn.

Table 33 'a' Prefixes used in replacement nomenclature

Element	'a' prefix	Element	'a' prefix	Element	'a' prefix
F	fluora	Au	aura	Eu	europa
Cl	chlora	Ni	nickela	Gd	gadolina
Br	broma	Pd	pallada	Tb	terba
I	ioda	Pt	platina	Dy	dysprosa
At	astata	Co	cobalta	Ho	holma
O	oxa	Rh	rhoda	Er	erba
S	thia	Ir	irida	Tm	thula
Se	selena	Fe	ferra	Yb	ytterba
Te	tellura	Ru	ruthena	Lu	luteta
Po	polona	Os	osma	Ac	actina
N	aza	Mn	mangana	Th	thora
P	phospha	Tc	techneta	Pa	protactina
As	arsa	Re	rhena	U	urana
Sb	stiba	Cr	chroma	Np	neptuna
Bi	bisma	Mo	molybda	Pu	plutona
C	carba	W	tungsta[a]	Am	america
Si	sila	V	vanada	Cm	cura
Ge	germa	Nb	nioba	Bk	berkela
Sn	stanna	Ta	tantala	Cf	californa
Pb	plumba	Ti	titana	Es	einsteina
B	bora	Zr	zircona	Fm	ferma
Al	alumina	Hf	hafna	Md	mendeleva
Ga	galla	Sc	scanda	No	nobela
In	inda	Y	yttra	Lr	lawrenca
Tl	thalla	La	lanthana	Be	berylla
Zn	zinca	Ce	cera	Mg	magnesa
Cd	cadma	Pr	praseodyma	Ca	calca
Hg	mercura	Nd	neodyma	Sr	stronta
Cu	cupra	Pm	prometha	Ba	bara
Ag	argenta	Sm	samara	Ra	rada

[a]Also wolframa

[130] Specific names for the two forms of metheno would be methanylylidene and methanetriyl, respectively.
[131] Specific names for the two forms of nitrilo would be azanylylidene and azanetriyl, respectively.

Index

An initial or a terminal hyphen indicates that the syllables occur only at the end or at the beginning, respectively, of the names of a compound.

abeo- 31
Absolute configuration 152
Aceanthrylene 164
Acenaphthoquinone 174
Acenaphthylene 164
Acephenanthrylene 164
Acetals 102
Acetic acid 175
Acetone 174
Acetoacetic acid 176
Acetonyl 174
Acetophenone 174
Acetylene 162
Acid halides 63, 122, 175
Acids
  as principal group 70
  aldehydic 107, 109
  amic 107, 109
  amino 110
  anilic 107, 109
  hydrazonic 110, 111
  hydroxamic 111
  hydroximic 110, 111
  hydroxy 109
  imidic 110, 111
  oxo 109
  peroxy 110
  thioic 111
Acridarsine 166
Acridine 166
Acrylic acid 175
Acyl groups 175
Acylals 102, 103
Acylamino 127
Acyloins 104
Adamantane 166
Added hydrogen (see Indicated hydrogen)
Additive name 16
Additive operation 24, 25
-adiene 59
-adienyne 60
Adipic acid 176
-adiyne 59
-al 63, 98
Alcoholates 63
Alcohols 63, 91
  as principal group 70
Aldehydes 63, 98, 175
  as principal group 70
Aldimines 89, 90
Aldoximes 104
-alene 47
Allene 162
Allophanic acid 179

Allophanyl 179
Alloxane 177
Allyl 163
-amide 63, 107
Amides 175
  as principal group 70
Amidines 63
Amido 66
Amidyl 135
-amine 63
Amine oxides 90
Amines 125–127
  primary 88
  as principal group 70
  secondary 88, 89
  tertiary 88, 89
Amino 63
Aminocarbimidoyl 63
Aminocarbonyl 63
Aminyl 135
Aminylene 132
Amminium 136
Anhydrides 123
  chalcogen analogues of 124
  as principal group 70
  symmetrical 123
  unsymmetrical (mixed) 124
Anhydrosulfide 123
Anilides 126
Aniline 177
Anilino 177
Anions 139–141
  affixes for 64
  as principal group 70
Anisole 173
[*n*]Annulenes 39, 45, 61
Anthracene 164
Anthranilic acid 176
Anthraquinone 174
Anthryl 166
Antimony
  organometallic compounds of 81
  pentavalent 78
  tetravalent 78
-aphene 47
Arsane 37, 78
Arsanthridine 166
Arsenic
  pentavalent 78
  tetravalent 78
Arsenic acids 115, 116
Arsindole 166
Arsine 78
Arsinic acid 116

Arsinimidic acid   116
Arsinoline   167
Arsonic acids   115, 116
  as principal group   70
Arsonium   136
Arsorane   79
Assemblies of identical units   32, 33, 34, 53
Assembly name   16
Astatane   37
-ate   64, 139, 140
-ato   64, 140
-atriene   59
-atriyne   59
Azane   37
Azanyl   132
Azanylia   138
Azides   87
Azido   66, 69
Azines   105
Azinic acid   65
Azino   105
Azinoyl   65, 90
Azo compounds   83, 84, 86
Azonia   138
Azonic acid   65
Azono   65
Azonyl   65
Azoxy compounds   83, 84, 86
Azulene   164

Barbituric acid   177
Benzene   162
Benzeno   181
Benzhydryl   163
Benzidine   177
Benzidino   177
Benzil   174
Benzilic acid   176
Benzoic acid   175
Benzophenone   174
Benzoquinone   174
Benzyl   163
Benzylidene   163
Biacetyl   174
Bicyclo   49
Biguanide   179
Biphenyl   54
Biphenylene   48
Bipyridyl   54
Bismuth
  organometallic compounds of   81
  pentavalent   78
  tetravalent   78
Bismuthane   37, 78
Bismuthine   78
Bismuthonium   136
Bismuthonia   138
Biuret   179
Bond migration   31
Bonding number   17, 20, 21
Borane   37
Boranylia   138
Boron hydrides   37
Braces   7
Brackets   6
Bromane   37
Bromo   66, 69, 81
Bromoform   180
Bromonium   136

Butadieno   180
Butane   162
Buteno   180
Butano   180
Butoxy   173
sec-Butoxy   173
tert-Butoxy   173
sec-Butyl   163
tert-Butyl   163
Butyric acid   175

c (abbreviation for cis)   150
Carbaldehyde   98
-carbaldehyde   63
Carbamic acid   176
Carbamimidoyl   63
Carbamoyl   63, 125
Carbazole   167
Carbazone   179
Carbozono   179
Carbene   132
Carbodiazone   179
Carbodiimide   179
Carbohydrazide   128
Carbohydrazonic acid   111
Carbohydroxamic acid   111, 112
Carbohydroximic acid   111
Carbolactone   107, 120
β-Carboline   167
Carbonitrile   130
-carbonitrile   63, 107
Carbonohydrazide   179
-carbonyl halide   63
Carbonyl halides   107
Carbonylium   138
Carboselenaldehyde   99
Carboselenothioic acid   112
Carbothialdehyde   99
Carbothioic acids   111
Carboxamide   107, 125
-carboxamide   63
Carboxamido   127
Carboxanilide   126
-carboximidamide   63
Carboximidic acid   111
Carboxy   63
-carboxylate   63
Carboxylates   63
Carboxylato   63, 117
-carboxylic acid   63
Carboxylic acids   63, 107
Carboxylic anhydrides   107
Carvacrol   173
Cations   136–139
  affixes for   64
  as principal group   70
Chalcogen hydrides   78
Chalcone   174
Characteristic group   13, 20, 25, 26, 62
Chlorane   37
Chloro   66, 69, 82
Chloroform   180
Chloronium   136
Chlorosyl   69
Chloryl   69
Chromane   171
Chromene   167
Chrysene   164
Cinnamic acid   176

INDEX

Cinnamyl 163
Cinnoline 167
Citric acid 176
Configuration 149
  absolute 152
  relative 154
Conjunctive name 15
Conjunctive operation 26
λ-Convention 141
Coordination nomenclature 25
Coronene 164
Cresol 173
Cubane 166
Cumene 163
Cyanate 129, 131
Cyanato 66, 131
Cyanide 129, 130, 131
Cyano 63, 66, 130, 131
Cyclo 30, 38, 44
Cycloalkanes 39
Cyclophanes 55
Cymene 163

Dehydro prefixes 61, 69
Diacylsulfanes 124
-dial 98
Diazane 129
Diazano 181
Diazenes 83, 84, 86
Diazeno 181
Diazo compounds 83, 87
Diazonium compounds 87
Dicarboximide 107
Dicarboxylic acids 107
Didehydro- 61
Dihalo-λ^3-iodanyl 69
Dihaloiodo (abandoned) 69
Dihydronitroryl 65
Dihydrophosphoryl 65
Dihydroxy-λ^3-iodanyl 69
Dihydroxyiodo (abandoned) 69
Dihydroxynitroryl 65
Dihydroxyphosphanyl 65
Dihydroxyphosphoryl 65
-diide 139
Diium 136
Dimethoxyphosphanyl 65
Dioxidane 97
-dioic acid 107
-diol 91
-dione 100
Diphosphanes 78, 79
  as principal group 70
Diselenides
  as principal group 70
Disiloxane 79
Disulfides
  as principal group 70
Ditellurides
  as principal group 70
Ditelluroacetals 102
Ditellurohemiacetals 103
Dithioacetals 102
Dithiohemiacetals 103
Dithioperoxy 66
Dithiosulfonic acid 114
-diyl 56
-diylidene 56
-diylidyne 56

Enclosing marks 5
-ene 59, 62
-enediyne 60
-enyne 60
Epicyclopenta 180
Epicyclopropa 180
Epidioxy 181
Epidithio 181
Epimino 181
Episeleno 181
Epithio 181
Epoxy 95, 181
Epoxyethano 181
Epoxyimino 181
Epoxymethano 181
Epoxypropeno 181
Epoxythio 181
Epoxythioxy 181
Esters 117, 118
  as principal group 70
Ethoxy 173
Ethane 162
Ethano 180
Etheno 180
Ethers 63, 94, 95
  cyclic 95
  as principal group 70
Ethoxy 173
Ethylene 57, 163
Ethylene glycol 173
Ethylenediaminetetraacetic acid 176

Fluorane 37
Fluoranthene 164
Fluorene 164
Fluoro 66, 69, 82
Fluoroform 180
Fluoronium 136
Formazan 179
Formic acid 175
Formyl 63, 98
Fulfminate 131
Fulminato 131
Fulvene 163
Fumaric acid 175
Functional class name 14, 16, 20
Functional modifiers 64
Functional parent 13, 59, 65
Functional replacement 66
Furan 167
Furano 181
Furfuryl 172
Furoic acid 175
Furyl 172
Fusion nomenclature 44

Germane 37
Germanium
  organometallic compounds of 81
Glutaric acid 176
Glyceric acid 176
Glycerol 173
Glycolic acid 176
Glyoxal 176
Glyoxylic acid 176
Guanidine 179
Guanidino 179

Halocarbonyl 63
Haloformyl (abandoned) 63
Hantzsch–Widman name 14, 37, 40, 42, 45, 49, 59
-helicene 48
Hemiacetals 102, 103
Heptalene 47
Heteroatomic bridge 181
Heterocyclic bridges 181
Heteromonocycles 43
Hexahelicene 48
Hexaphene 47
Hydantoic acid 179
Hydantoin 177
Hydantoyl 179
Hydazones 105
Hydrazides 78, 128, 12, 175
  as principal group 70
Hydrazido 66
Hydrazines 128, 179
  as principal group 70
Hydrazido 66
Hydrazino 179
Hydrazonic acids 110, 111
Hydrazono 105
Hydrido 82
Hydro 60, 69
Hydrocarbon bridge 180
Hydrocarbons 36, 39, 77
Hydrohydroxynitroryl 65
Hydrohydroxyphosphoryl 65
Hydron 94
Hydronitroryl 65
Hydroperoxides 63, 96
  as principal group 70
Hydroperoxy 63, 96
Hydrophosphoryl 65
Hydropolyselenides 97
Hydropolysulfides 97
Hydropolytellurides 97
Hydroquinone 173
Hydroseleno 93
Hydroxamic acids 111, 112
Hydroximic acids 110, 111
Hydroxy 63, 92
  compounds 81–98
Hydroxy acids 109
Hydroxyimino 104
Hydroxylamines 90
Hydroxynitroryl 65, 69
Hydroxyoxidophosphanyl 65
Hydroxyphosphanediyl 65
Hydroxyphosphanylidene 65
Hydroxyphosphoryl 65

-ida 64, 141
-ide 64, 139, 141
-idyl 64
Indicated hydrogen 60
  definition 34
-imidamide 63
Imidazole 167
Imidazolidine 171
Imides 107, 128, 175
  as principal group 70
Imidic acids 110, 111
Imido 66
-imine 63
Imines 63, 78, 87, 89

as principal group 70
Imino 63, 134
Iminyl 135
as-Indacene 164
s-Indacene 164
Indazole 167
Indene 164
-io 64, 138
Iodane 37
Iodo 66, 69, 82
Iodoform 180
Indane 166
Indole 167
Indoline 171
Indolizine 167
Iodonium 136
Iodosyl 69
Iodoxy (abandoned) 69
Iodyl 69
Ions 131
Isoarsindole 167
Isoarsinoline 167
Isobenzofuran 167
Isobutane 163
Isobutoxy 173
Isobutyl 163
Isobutyric acid 176
Isochromane 171
Isochromene 167
Isocyanate 131
Isocyanato 66, 131
Isocyanides 129, 131
Isocyano 66, 131
Isodiazenes 87
Isoindole 168
Isoindoline 171
Isomerism
  cis 149, 150
  (*E*) 149, 151, 152
  trans 149, 150
  (*Z*) 149, 151, 152
Isonicotinic acid 175
Isopentane 163
Isopentyl 163
Isophosphindole 168
Isophosphinoline 168
Isophthalic acid 175
Isoprene 163
Isopropenyl 163
Isopropoxy 173
Isopropyl 163
Isopropylidene 163
Isoquinoline 168
Isoquinolone 174
Isoquinolyl 172
Isoselenocyanate 131
Isoselenocyanato 131
Isothiazole 168
Isothiocyanate 131
Isothiocyanato 66, 131
Isotopically deficient compounds 160
Isotopically labelled compounds 157
Isotopically modified compounds 155–161
Isotopically substituted compounds 156
Isourea 179
Isoxazole 168
Italicization 7
-ite 64, 139
-ium 64, 136, 141

186

Ketals   102
Ketenes   101, 174
Ketimines   89, 90
Ketones   63, 100
  as principal group   70
Ketoximes   104

Lactams   119, 120
Lactic acid   176
Lactims   119, 120
Lactones   119, 120
Lead
  organometallic compounds of   81
Locants   17
  lowest set   17
  position   1

Maleic acid   175
Malonic acid   175
Mercapto   63, 93
Mesityl   163
Mesitylene   163
Mesyl   178
Methacrylic acid   176
Methane   37, 162
Methano   180
Methoxy   93, 173
Methoxyhydroxyphosphoryl   65
Methylene   57, 132
Monoselenoacetals   102
Monoselenohemiacetals   103
Monothioacetals   102
Monothiohemiacetals   103
Morpholine   171
Morpholino   57
Multiplicative operation   32
Multiplying affixes   71

Name construction   20, 22, 68
Name interpretation   143–148
Naphthalene   164
Naphthaleno   181
Naphthoic acid   175
Naphthone   101
Naphthoquinone   174
Naphthyl   166
-naphthylene   48
Naphthyridine   168
Natural product hydrides   55
Neopentane   163
Neopentyl   163
Nicotinic acid   175
Nitrene   132
Nitrido   66
-nitrile   63, 107
Nitrile oxides   131
Nitriles   63, 129–131, 175
Nitro   69, 83
aci-Nitro   83
Nitrogen hydrides   78
Nitroryl   65
Nitroso   69, 83
Nomenclature operations   19, 22
Nonselectively labelled compounds   160
Nuclide symbols   155
Numerical (multiplicative) prefixes   4

-oate   63, 107
Octalene   47
-ohydrazide   128
-oic acid   63, 107, 108
-oic anhydride   107
-ol   63, 91, 140
-olactone   107, 120
-olate   63
Oleic acid   176
-one   63, 100
-onia   64, 138
-onio   64, 138
-onium   64, 136
Organometallic compounds   81, 82
ortho-fused polycycles   44, 46
Ovalene   164
Oxalic acid   176
Oxamic acid   176
Oxamide   176
Oxidane   37, 94
Oxides   85, 95
Oxido   63
Oximes   104
Oxo   63, 98, 134
Oxo acids   109
Oxonia   138
Oxonium   136
Oxy   63, 69
-oxycarbonyl   63
Oxyl   135
-oyl halide   63

Palmitic acid   176
Parent hydrides
  acyclic polynuclear   36
  bridged   49
  definition   13
  monocyclic   39
  mononuclear   36, 37
  natural product   55
  polycyclic   44
  polyspiro   52
  spiro   51, 53
Parent structure   13, 18, 19, 20, 162
Parentheses   5, 6
Pentacene   47
Pentaerythritol   173
Pentalene   47
Pentaphene   47
tert-Pentyl   163
peri-fused polycycles   44, 46
Perimidine   168
Peroxides   63, 96
  as principal group   79
Peroxy   63, 66, 96
Peroxy acids   110, 111
Peroxycarboxylic acids   111
Peroxyl   135
Peracetic acid   176
Perbenzoic acid   176
Performic acid   176
Perylene   165
Phenacyl   174
Phenalene   165
Phenanthrene   165
Phenanthridine   168
Phenanthroline   168
Phenanthryl   166

Phenazine 168
Phenethyl 163
Pheno- 49
Phenol 173
Phenolates 63
Phenols 63, 91
   as principal group 70
Phenone 101
10*H*-Phenoselenazine 49
Phenoxathiine 49
Phenoxy 93, 173
Phenyl 163
-phenylene 47, 163
Phenylene 57
Phosgene 180
Phosphanes 37, 78
   as principal group 70
Phosphano 181
Phosphinato 65
Phosphindole 168
Phosphine 37, 78
Phosphinic acid 65, 116
   as principal group 70
Phosphinimidic acid 116
Phosphinimidothioic acid 116
Phosphinodithioic acid 116
Phosphinoline 168
Phosphinothioic acid 116
Phosphinous acid 65
Phosphonato 65, 117
Phosphonic acid 65, 116
Phosphonium 136
Phosphono 65
Phosphonodithioic acid 116
Phosphonothioic acid 116
Phosphonotrithioic acid 116
Phosphonoyl 65
Phosphorane 37, 79
Phosphorous acid 65
Phosphorus
   pentavalent 78
   tetravalent 78
Phosphorus acid 115
Phosphoryl 65
Phthalazine 168
Phthalic acid 175
Picene 165
Picric acid 173
Pinacol 173
Piperazine 171
Piperidine 171
Piperidino 58
Piperidyl 172
Pleiadene 165
Plumbane 37
Polane 37
Polyethers 95
Polyselane 97
Polyselenides 97
Polyspiro parent hydride 52
Polysulfides 95
Polytellane 97
Polytellurides 97
Polyvalent bridges 182
Prefixes
   additive 12, 24
   detachable 19
   non-detachable 10
   numerical (multiplying) 4

   order of 10
   subtractive 12
Principal component 45
Principal group 26
   definition 13
   priority of 70
Priority of principal characteristic
   groups 69
Prismane 166
Propane 162
Propano 180
Propeno 180
Propiolic acid 175
Propionic acid 175
Propiophenone 174
Propoxy 173
Pteridine 168
Punctuation 2
Purine 168
Pyran 169
Pyranthrene 165
Pyrazine 169
Pyrazole 169
Pyrazolidine 171
Pyrene 165
Pyridazine 169
Pyridine 169
Pyridyl 172
Pyrimidine 169
Pyrocatechol 173
Pyrrole 169
Pyrrolidine 171
Pyrrolidone 174
Pyrrolizine 169
Pyruvic acid 176
Pyrylium 172

Quaterphenyl 55
Quinazoline 169
Quinoline 169
Quinolizine 169
Quinolone 174
Quinolyl 172
-quinone 100
Quinoxaline 169
Quinuclidine 171

r (abbreviation for reference substituent) 150
(*R*)-Configuration 152–154
Radical centers 134
Radical ions 141, 142
Radicals 132–136
   affixes for 64
   bivalent 132
   monovalent 132
   as principal group 70
   trivalent 132
Radicofunctional name 15
Rearrangement 31
rel- (abbreviation for *relative*) 154
Relative configuration 154
Replacement name 15
Replacement nomenclature 51
Replacement operation 23
Resorcinol 173
retro- 32
Rhodanine 177
Ring assemblies 53
Ring cleavage 30

Ring formation   30
Rubicene   165

(S)-Configuration   152–154
Salt name   16
Salts   63, 117
    of acids   63
    of alcohols   94
Seco-   31
Selane   37, 94
Sclanyl   69, 92
Selectively labelled compounds   158
Selenal   99
Selenides   94, 95
    as principal group   70
Seleninyl   97
Selenium acids   116
Seleno   66, 91, 99, 101
Selenoaldehydes
    as principal group   70
Selenocarbaldehyde   99
Selenocyanate   131
Selenocyanato   131
Selenohydroperoxides
    as principal group   70
Selenoic acid   112
Selenoketones
    as principal group   70
Selenols   91, 103
    as principal group   70
Selenones   97
Selenonic acids
    as principal group   70
Selenonium   136
Selenonyl   97
Selenophene   169
Selenosemicarbazide   105
Selenothioacetals   102
Selenothiohemiacetals   109
Selenoxides   97
Selenoxo   101
Selone   101
Semicarbazide   105, 179
Semicarbazido   179
Semicarbazone   105
Semicarbazono   105
Semisystematic name   14, 162
Semitrivial name   14
Seniority
    definition   17
Silane   37, 79
Silaselenanes   79, 80
Silatelluranes   79, 80
Silathianes   79, 80
Silazanes   79, 80
Siloxanes   79, 80
Silylene   132
Spiro parent hydrides   51
Stannane   37
Stannano   181
Stearic acid   176
Stereochemical specification   149–154
Stereodescriptors   18, 149
Stereoparents   18
Stibane   37, 78
Stibine   78
Stibonium   136
Stilbene   163
Styrene   163

Styryl   163
Substitutive name   15
Substitutive operation   22
Subtractive name   16
Subtractive operation   27
Succinic acid   175
Suffix   63
    additive   25
    denoting multiple bonds   59
Sulfane   37, 94, 97, 124
Sulfanilic acid   178
λ^4-Sulfano   181
Sulfanyl   63, 69, 92, 132
Sulfides   94, 95
    as principal group   70
Sulfido   63
Sulfinic acids   113, 114
    as principal group   70
Sulfinimidic acid   114
Sulfino   113
Sulfinohydrazonic acid   114
Sulfinohydroxamic acid   114
Sulfo   63, 113
Sulfonamides   126
-sulfonate   63
Sulfonates   63
Sulfonato   63, 117
Sulfones   97
-sulfonic acid   63, 113
Sulfonic acids   63
    as principal group   70
Sulfonimidic acid   114
Sulfonium   136
Sulfonodiimidic acid   114
Sulfonohydrazonic acid   114
Sulfonohydroxamic acid   114
Sulfoxides   97
Sulfur acids   113
Sultams   121
Sultones   114, 120

t (abbreviation for trans)   150
Tartaric acid   176
Tellane   37, 94
Tellanyl   69, 92
Tellurides   94, 95
    as principal group   70
Telluro   66, 91, 99
Telluroaldehydes
    as principal group   70
Tellurohydroperoxides
    as principal group   70
Telluroketones
    as principal group   70
Tellurols   31
    as principal group   70
Telluronium   136
Tellurophene   169
Terephthalic acid   175
Terphenyl   54
Tetracyclo   50
Tetraguanide   179
Tetranaphthylene   48
Tetrauret   179
-tetrayl   56
Thenyl   172
Thial   99
Thianthrene   49
Thienyl   172

Thio   66, 91, 99, 101
Thioaldehydes   98
    as principal group   70
Thioanhydride   124
Thiocarbaldehyde   99
Thiocarbonic acids   112
Thiocarboxylic acids   112
Thiocyanate   131
Thiocyanato   66, 131
Thioformyl   99
Thiohydroperoxides
    as principal group   70
Thioic acids   111
Thioketones
    as principal group   70
-thiol   63, 91
Thiols   63, 103, 140
    as principal group   70
-thiolate   63
Thiolates   63
Thione   101
Thionia   138
Thioperoxy   66
Thiophene   169
Thiosemicarbazone   105
Thiosphosgene   180
Thiosulfinic acid   114
Thiosulfonic acid   114
Thiosulfonimidic acid   114
Thioxo   99, 101, 134
Thiuram disulfide   178
Thiuram monosulfide   178
Thymol   173
Tin
    organometallic compounds of   81
Toluene   163
Toluidine   177
Toluidino   177
Tolyl   163
Tosyl   178
Triazeno   181

Tricyclo   50
Triguanide   179
-triide   139
Trimethylene   57
Trinaphthylene   48
Trisiloxane   79
Trithiosulfonic acid   114
Trityl   163
Triuret   179
Trivial name   14, 162
-triyl   56

-uida   64, 141
-uide   64, 139, 141
-uidyl   64
Unsaturation   59
Urea   179
Ureido   179
Ureylene   179

Vinyl   163
von Baeyer system for bridged rings   49

Xanthene   169
Xylene   163

-yl   56, 64
-ylene   57
-ylia   64, 138
-ylidene   56, 133, 138, 141, 162
-ylidyne   56, 133, 162
-ylium   64, 138, 141
-yliumyl   64
Ylo-   134
-ylylidene   56
-ylylidyne   56
-yne   59

Zwitterionic compounds   141
    as principal group   70